THE
BRAZER'S
HANDBOOK

THE BRAZER'S HANDBOOK

Charles R. Self, Jr.

🏠 **A DRAKE HOME CRAFTSMAN'S BOOK**

Drake Publishers Inc. New York

Published in 1978 by
Drake Publishers, Inc.
801 Second Avenue
New York, N.Y. 10017

Library of Congress Cataloging in Publication Data

   Self, Charles R., Jr.
   The brazer's handbook.

   (Drake Home Craftsman Series)
   1. Brazing. I. Title.
TT267.S44   671.5'6   78-56960
ISBN 0-8473-1759-5

Printed in the United States of America

CONTENTS

THE BRAZER'S HANDBOOK

PREFACE

For many of us who work at home or on the farm, trying to keep repair bills within reason, or to minimize down time of equipment, metal joints can be a real problem. Often a joint that really requires welding for strength is made by tossing in a screw or bolt. Or an attempted repair to an air conditioner may be short lived because soft soldering was used when the vibration of the unit necessitated hard soldering. While fusion welding, whether with oxy-fuel gas equipment or electric arc gear, is somewhat difficult to learn, and to remember when not applied rather often, there are two other techniques — brazing and braze welding — that will aid almost any of us a great deal. Both are rather simple to learn. And in addition to their simplicity, both have some built-in features that serve as failsafe devices, so that great skill in the processes results, generally, in little more than extremely neat, attractive joints, whereas lesser skill results in uglier joints having about the same strength.

Fortunately for most of us, the equipment needed for forming these joints is seldom as expensive as it is thought to be. For example, braze welding of lighter metals that don't transfer heat at an excessive rate can often be carried out with torches costing no more than $20.

This book covers the more sophisticated soldering techniques, known as silver or hard soldering (also called brazing), as well as the more difficult braze welding techniques. Braze welding will often require more expensive equipment, and does require the melting of the base metals being welded for proper joint construction. Braze welding offers greater ease of repair in such metals as cast iron and aluminum than does fusion welding, while still giving very great joint strength. At the same time, braze welding offers a chance to join dissimilar metals, something that just cannot be done by fusion welding, where the alloys of the parts being joined must be quite similar if the weld is to be successful.

Simple soldering isn't covered in this book for a simple reason. The difficulties present in the use of soft solders are so easily overcome, the techniques so readily available, that there is little or no reason to include a great deal of text on the subject. Almost any soldering iron or propane torch comes with enough instruction to make basic soft soldering a simple job.

Silver soldering, though, offers a field of greater interest for everything from belt buckle repair (many of the fancier new belt buckles are put together this way) to jewelry making. This requires higher heat levels (over 427°C. — 800°F.), which in turn demands a different level of equipment. Joint strength is correspondingly greater with silver soldering, or brazing, too.

Both braze welding and brazing depend strongly on the tinning effect of the first thin coat of filler metal to form a molecular bond with the base metal(s) being joined. Anything that interferes with this bond weakens the resulting joint, often quite badly. Therefore the mechanical and chemical cleaning processes must be carried out with an eye to perfection. The heat used, although high in terms of the possibility of burning skin, is low in relation to the melting points of most metal oxides and other contaminants that would often disappear in fusion welding. Fluxing and a good mechanical cleaning take care of such things with these simpler processes.

You will note throughout the book a strong emphasis on safety. Such an emphasis is essential, for in all the work described here you will be working with high temperatures and, usually, volatile gases. A misstep will result in burns or possible explosions or fires. Essential to the safety process is the proper equipment in top condition: welding gloves; asbestos millboard; firebrick; fire extinguishers; gases held in proper containers, properly stored and handled; hoses and torches in good shape. Remember at all times while working with even propane torches that the temperature of the flame is up around 4500°F. when combined with pure oxygen, and that acetylene will provide a flame temperature nearly 1100 degrees hotter! Not the sort of thing you would care to brush across your clothing or, worse, bare skin.

You will also find strong emphasis here on the use of MAPP gas or its counterparts sold under other brand names: MAPP, as is Bernzomatic's Clean Burn high temperature fuel, is a stabilized form of propane, methylacetylene propadiene. It offers a flame temperature, in oxygen, only 300 degrees lower than basic acetylene, but it is easier to transport and use because it is less sensitive to shock than acetylene, has low toxicity (and high stink so that leaks are easy to trace), and has much lower explosive limits in air. MAPP is explosive, in air, within a 3.4 percent to 10.8 percent range, while acetylene is explosive in air from 2.5 percent to 80 percent concentrations.

For these reasons, and others that will come up as you progress through the book, my recommendation for all amateur brazers, and most professionals, would be a switch to MAPP gas in any area where acetylene is now used, or is planned for use. Brazing doesn't require the extra heat of acetylene — few welding processes do, either — so take safety to the top of your list.

Chapter 1

SAFETY

Safety when using open flames and great heat is of paramount importance. The hazards are many, but are reducible to a minimum. Almost any brazing job can be carried out with little fear of explosion, burn, or fire if proper methods of set-up and working are followed.

As a start, a check of the surface on which the work will be carried out is needed. Some heat joining methods can be carried out on the kitchen table, but both brazing and braze welding use heat of a high enough degree to make this unsafe without special precautions. Too, no torch used outdoors is completely safe unless the surface under the pieces being joined is checked carefully. Indoor work should take place on firebrick or asbestos surfaces only. Asbestos millboard is moderately expensive if you are only making a few joints, so I would recommend for small jobs the purchase of five or six firebricks for use as a working surface. Generally, the firebrick will provide a completely safe working spot. (Fig. 1-1.)

If work is being carried out in a home shop, it is easily possible to set up a permanent or semi-permanent firebrick covered bench, with thin asbestos millboard used to cut off heat access to any flammable materials (for this sort of heat cut-off, quarter-inch-thick millboard is sufficient, so that the expense is reduced). There are also nonflammable curtains available from most welding supply houses. These make the workspace portable, but also tend to run expenses rather high.

Working with high heat torches outdoors provides a need for a work surface, just as does working indoors. Too many novice braze welders will plop their metalwork right on a chunk of rock and go to heating things up. If the heat is applied for any extensive period of time, there is every possibility that the rock used as a workbench will explode. Any sort of rock, brick, or non-heat-resisting stone should be avoided when brazing or braze welding. Again, use asbestos millboard or firebrick as a work surface, or

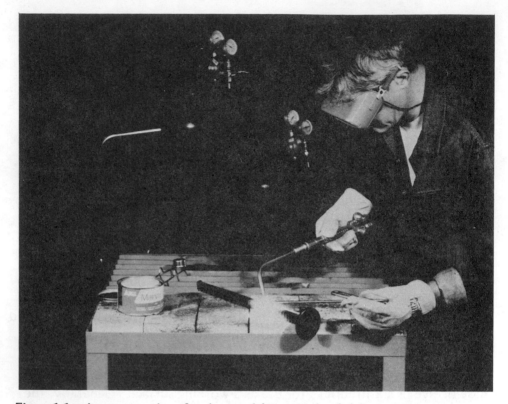

Figure 1-1. A proper work surface is essential to a good, safe job.

set the work up so that you are not forced to bring the flame in contact with anything other than the work and the atmosphere. (Fig. 1-2.)

Many of the precautions necessary with gas equipment for brazing and braze welding are simple, common-sense measures for keeping heat and flame away from combustible materials. Start with your own clothing. Make sure that any synthetic fabrics worn are not of a kind that melts or flames easily if touched by a torch flame or a piece of spattering hot metal. Use a good quality welding glove once you go beyond the basic brazing process into braze welding, where temperatures of both the torch flame and metals being joined are likely to be a thousand degrees hotter. Such gloves are available from many sources, and can be had in heat-resistant

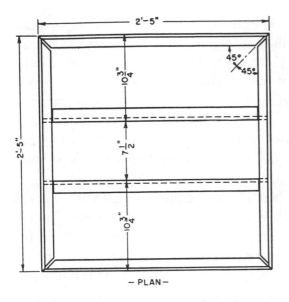

– PLAN –

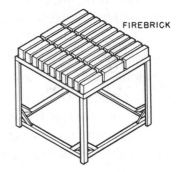

ISOMETRIC VIEW OF WELDING TABLE

Material Required for a Welding Table

33 Firebricks
36 Linear Feet of
   1½" x 1½" x 1¼"
   Angle
   Iron 2.34 lb. per foot
   Note: All joints to be
         welded.
   Firebrick 2¼" x 4½" x 9"

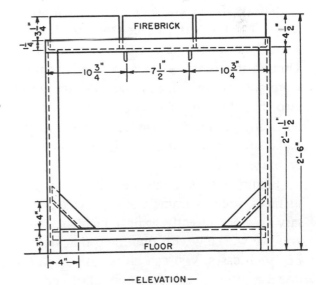

—ELEVATION—

Figure 1-2. A good project for the beginning braze welder is a safety bench. AIRCO

fabric or heat-resistant leather. The cotton terry gloves are cheaper than the leather, but the leather will generally outwear fabric styles in heavy use. Wear leather footwear; spattering metal can sometimes hit a pair of sneakers and cause bad foot burns. Wear long sleeves. Don't work with loose clothing on, no matter how comfortable that long scarf may be on a cold day.

Check for the possible fall of molten filler material: if your body is likely to be underneath the fall, either relocate the work to prevent this, or buy protective clothing.

After you are suitably clothed, check the area in which the work is to be done. Look for flammables of all kinds: paint, thinner, gasoline, kerosene, fuel oil — anything that might be exploded or set aflame by a flying piece of hot metal or a spark, or even the torch flame in the case of volatile gases or liquids such as gasoline or propane. If the item, such as a basement fuel oil tank, cannot be relocated, move the work. In some cases you will be easily able to isolate the flammable material or the work, whereas in others difficulty will be involved in moving or covering either.

Safety always comes first, so do whatever is needed to prevent problems. To start with, never store gasoline in the house. Most of us keep a small can in the garage or woodshed for use with various power tools. Though this is still a poor idea, it is sometimes hard to avoid. I now store any gasoline needed outside the house, in cans, covered with a tarp, but this isn't always possible in urban areas where both the contents and cans may disappear as soon as the owner's back is turned. Make sure all such material is moved outdoors before starting to braze or braze weld.

Flammable materials are only a part of the problem. Combustible surfaces, such as frame walls, provide another part. It is never necessary to work directly against such surfaces, in my experience, unless you are working with plumbing that has to be sweat soldered into place. For such cases, you can either use a soldering iron instead of a torch or insert a piece of asbestos millboard between the work and the combustible surface.

With the danger of fire from heat or flame reduced as far as possible, you have reached the point where you check the equipment set-up and treatment to be sure that the explosive gases you are using will be even less than a minor hazard. In many cases, only a single gas torch will be needed to carry out the jobs covered in this book, but for actual braze welding of heavier materials, you must use a fuel gas, usually acetylene or MAPP, and oxygen, piped into the torch through hoses from individual tanks. The set-up will vary. Some small outfits do not offer hoses, but have small tanks attached directly to the torch head (Bernzomatic), while others use only a single tank, with oxygen provided by a burning "candle" (Cleanweld's Solidox). Other small, but near professional quality, equipment offers hoses, torch heads designed for various jobs, small tanks, and greater versatility. The variation in price is great, though, as a unit such as Airco's Tote-Weld, with a 20-cubic-foot refillable oxygen cylinder and a throwaway MAPP cylinder, can approach $165, complete, whereas the smaller torches seldom exceed $35. Of course, capacity and ease of use are greater, as is possible term of use for each job. The Solidox unit is limited to a tops of about twelve minutes of actual brazing time; the Bernzomatic runs possibly a few minutes longer. While I have no total time to give you for the Tote-Weld, I have now used mine, including a fair amount of cutting (which burns oxygen quite quickly) for over an hour, and the oxygen gauge shows a tank pressure of a bit over 600 pounds, something less than one-third the original 2000 psi or so.

The choice of equipment will be covered more thoroughly later, while we now look at the difference in the handling of the different tools. First, the single-gas torches, propane or MAPP. These require the fewest precautions, since pressures are only moderately high and the explosive qualities in air are rather low. MAPP gas is more expensive than propane, but it provides a hotter flame in both air and pure oxygen. MAPP is also insensitive to shock, and less explosive in air than is acetylene. (Acetylene is classified as unstable under impact shock conditions, and is explosive at percentages from 2.5 to 80 in air. MAPP is shock stable, and is explosive in air only from 3.4 to 10.8 percent. In addition, MAPP loses most of the tendency to backfire that acetylene has.) MAPP is cheaper than acetylene, and the cylinders are lighter. (Acetylene cylinders must have heavy built-in stabilizers, usually of porous cement and acetone.) For almost any job, with the exception of working with copper alloys over about 65 percent, MAPP is much to be preferred by the home brazer or welder.

Propane, for those silver soldering jobs, costs less than MAPP per cylinder, is listed as shock stable, and has explosive limits in air of 2.3 to 9.5 percent. The temperature of the flame is several hundred degrees lower, but this is not often a problem. In addition, propane torches are generally about one-half to one-third the cost of comparable single-gas MAPP torches because sophisticated torch-head designs are not often used to develop maximum heat. In any case, any single-gas torch designed for use with MAPP gas can be used with propane when money is more of a consideration than heat. In almost every case, propane torch heads can also be used with MAPP gas, but there is a loss of flame temperature because of the lack of a special head design.

BASIC SAFETY RECOMMENDATIONS

No matter the type of cylinder or the type of fuel gases, the use of *every* pressurized gas requires some primary safety precautions.

We've already covered number one, the removal of either the work or the material when flammables and combustibles are stored nearby. Number two is personal dress in appropriate heat and flame resistant clothing, and that too has been covered.

Number three is just as important, for the fumes turned out by metals under extreme heat, by fluxes, and by other chemicals that may be present, can be extremely noxious. Never weld where ventilation is inadequate. Some fumes may only make you sick to your stomach for a few minutes or a few hours; others, such as lead fumes, can actively poison you. If ventilation cannot be provided, either farm the job out or purchase adequate respirators.

Once heat is applied to the pieces being worked, keep your hands off them. Severe burns are possible. Even silver soldering requires temperatures of around 800° F. and more. And braze welding filler rod may not melt until temperatures exceed 1900° F. (Airco's 23A silicon copper rod melts at 1980° F.)

Pay attention to eye protection. Wear goggles when brazing and braze welding, of course, but also wear safety goggles when you're preparing the metal for the joints and when you finish the heat application and chip the slag from the surface. For braze welding, you will need goggles having a lens density listed as three or four, while hard soldering could use a two.

Take precautions with the cylinders of oxygen and fuel gas. No matter how insensitive to shock a fuel gas such as MAPP may be, you must remember that it is pressurized to 225 pounds per square inch. A crack in the cylinder, or a broken valve at the top, can cause quite severe problems ranging from a blown cylinder to a cylinder chasing you around the shop. With oxygen, which is pressurized to about 2000 psi, the hazards are intensified; smaller cracks can become major ruptures even more rapidly. And though oxygen itself is not flammable, it accelerates the combustion of other materials: a broken or leaking valve spewing oxygen on to a pile of oily rags can cause quite a conflagration.

In no case should oil *ever* be used on the working parts, threads, or joints of any fuel gas or oxygen hose. Oxygen in the presence of petroleum is a severe fire and explosion hazard. If lubrication is necessary, go to your welding supplier and have him recommend the product he is willing to use. I've yet to see a decently treated valve, gauge, or fitting that ever needed lubrication.

Always make sure the cylinders are stable, both during storage and during working periods. If possible, use a cylinder cart. If not, chain the cylinders to the workshop wall so that they cannot be tipped over. (Fig. 1-3.)

Always open cylinder valves with the appropriate tool, never with a hammer or a heavy wrench. Those valves are constructed of alloys of great strength, but they are not designed to take abuse.

If connections do not fit easily, re-check the set-up: you may just have the wrong hose on the fitting. Oxygen fittings will have right-hand connections; all fuel gas hoses and fittings will be left-hand threaded.

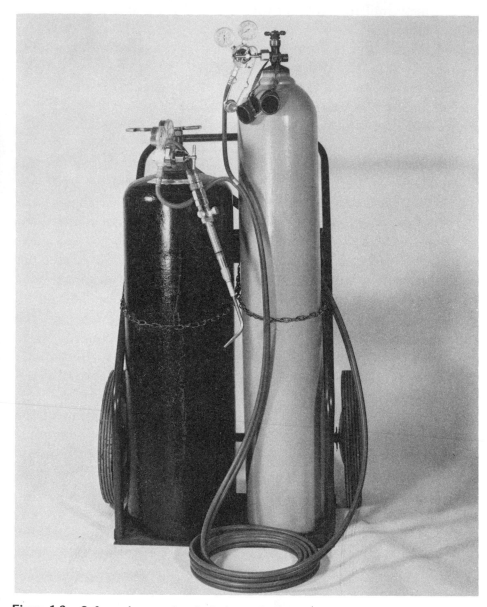

Figure 1-3. Safe work procedure includes tanks that cannot topple over. They should be chained to the cart or to a wall.

This, plus the different colors for the two types of hoses (oxygen will be green or black, the fuel hoses will be red), should keep anyone from getting them mixed up, but a few people always seem to feel the answer to any problem is a bigger wrench or hammer. In this case, a severe fire and explosion can easily result from forced fittings of the wrong kind, so use care, and never force a fitting that doesn't want to go.

Occasionally, a bit of dirt will be the problem when a fitting refuses to connect easily. Use a clean, *dry*, *grease-free* rag to wipe all fittings before assembling.

Keep all hoses up and away from areas where they might be run over by cars, lawnmowers, and other such vehicles. While the strong, nylon-braided neoprene hoses of quality oxy-fuel gas welding outfits will stand a lot of abuse, a single slit or crack can cause a lot of trouble. At the same time, keep the hoses free of oil and gas. Not only do these petroleum products provide an instant fire hazard around oxygen, but they also cause a deterioration of the rubber in the hoses, leading to later problems. Hoses should also be protected from flying slag, sparks, and so forth.

All fittings should be tight. Check them by brushing them with a solution of soapy water if you so much as suspect a leak. (Acetylene and MAPP gases both have distinctive odors; in the case of MAPP gas, the odor is an out-and-out stink, so the only leak that is hard to detect immediately is oxygen.)

Every torch kit comes complete with a spark-lighter. Use it. Matches provide nothing more than an excellent way to singe your hand when lighting the torch. If the torch blows out, close the valves and clear the hoses correctly before relighting the torch.

When work is stopped, release the pressure regulator valves by turning them to the left (during short work breaks). For longer breaks, such as overnight, shut the equipment down completely by first closing the tank valves, then opening the torch valves for a moment. Turn the pressure regulator valves to the left to release pressure on the diaphragms and prevent premature wear. (Fig. 1-4.)

1. Blow out cylinder valves before attaching regulators.

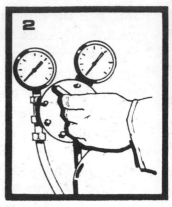

2. Release adjusting screw on regulators before opening cylinder valves.

Figure 1-4.   Basic safety rules for the use of oxy-fuel gas equipment.
DOCKSON CORPORATION

3. Stand to one side of regulator when opening cylinder valve.

4. Open cylinder valve SLOWLY.

5. Do not use or compress acetylene (in free state) at pressures higher than 15 PSI.

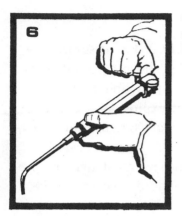

6. Purge oxygen and fuel gas passages (individually) before lighting torch.

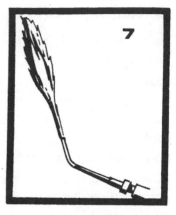

7. Light acetylene before opening oxygen valve on torch.

8. NEVER use oil on regulators, torches, fittings or other equipment in contact with oxygen.

9. Do not use OXYGEN as a substitute for compressed air.

10. Keep heat, flames and sparks away from combustibles.

CONTAINER BRAZING SAFETY

Safety when welding closed containers is an entirely different subject. Sealing cracks and holes in gas tanks, fuel oil tanks, and other containers that have at one time or another held volatile or combustible materials is a job that requires the utmost in care to prevent explosions. This is a field where experts are sometimes badly injured simply because they overlook a single, simple step in the cleaning-out process. To safely weld a container that has held a substance such as gasoline, we must first remove at least one of the three essentials to explosive reactions. These are the combustibles (in this case gasoline), a method of ignition, and the oxygen or air needed to support combustion. Eliminate any one of these and the job becomes safe. Obviously, working with the great heat of a torch, with its open flame, we cannot cut out the source of ignition. That leaves two areas for improvement, the removal of combustibles or the removal of the oxygen needed to support combustion. In almost every case, safety is enhanced if both procedures are followed. First, then, we must purge the container. An object such as an automobile fuel tank can be purged by steam cleaning thoroughly while all vents are left open. Purging can be done with inert gas, such as argon or nitrogen, if steam-cleaning equipment is not available. For gasoline, using inert gas, introduce the inert gas into the tank from the top, leaving a top vent open for the escape of combustibles. Should the combustibles be heavier than air, introduce the inert gas into the bottom of the container, so that the *air* will be forced out the top vents.

If steam cleaning is to be carried out, make sure the braze welding or brazing equipment is on hand and set up. Then open all vents in the container, and apply low-pressure steam for at least a half hour for the average twenty-gallon fuel tank. (Actual times are a minimum of one hour for every 200 gallons of container size, but it is best to exceed this whenever possible.) Next, light a small torch and attach it to the end of a long pole or rod. Get behind a barrier and move the torch across the open vents on the tank. Continue the flow of steam into the container while welding.

If the first weld doesn't seal the tank, repeat the *entire* purging process, *after* the tank is cooled to near room temperature.

For splits high up in fuel tanks, you can sometimes fill the tank with water so that any cracks are barely above the surface of the water. Keep the break at the top, and make sure that the water comes very, very close to filling the container or the danger of explosion will still be great. This is the *least* safe method of repairing fuel tanks, though it is often used for small vessels. In almost every case, I would prefer using an inert gas or steam purge over filling the container with water.

FUMES AS HAZARDS

At normal braze welding temperatures, several metals produce toxic fumes and others produce noxious fumes. In addition, many brazing and braze welding fluxes contain substances, such as fluorides, that may give off harmful vapors under high heat. Proper ventilation is essential, unless a respirator is worn.

Generally, the following metals, sans fluxes, produce few if any dangerous or irritating fumes: iron, aluminum, titanium, chromium, nickel, vanadium. Mild problems can occur with the following metal alloys: copper, zinc. The symptoms of poisoning by these two metals are similar, though less extreme with copper. These will include a headache, chills, and a rising and falling body temperature, along with a feeling of tightness in the chest. This is called metal fume fever and is accompanied by a severe case of nausea.

Severe problems can follow if these metals are heated in an area having improper ventilation: lead, manganese, cadmium. Lead oxides formed in heating are extremely soluble in human tissue; the effects of severe lead poisoning are cumulative over time. Symptoms include a "lead line" around the gums, a metallic taste in the mouth, constipation, vomiting, badly upset stomach. Manganese is not a common at-home metal, but may be found there once in a while. Manganese fumes can cause respiratory problems and varied unpleasant changes in the nervous system. Cadmium is likely to be found in plated objects. The fumes are deadly, and the area should be ventilated and a respirator worn at the same time. Cadmium's appearance is similar to that of mercury (bluish-white and shiny), which also gives off poisonous fumes.

Proper ventilation, then, is essential, even if you have to take things outdoors in mid-winter, or steal all the house fans in summer.

Fluxes are used to remove impurities on the metal and in the air when brazing, braze welding, soldering, or, with some types of metals, fusion welding. These chemical compounds differ in many ways, depending on the metals they are to be used on and the temperature ranges for which they are designed. (Obviously, soldering fluxes needn't be designed to hold up under braze welding temperatures, so they will be compounded differently.) Mild steel fusion welding seldom requires flux for a good job inasmuch as the heat needed to do the job, about $2800°$ F., effectively burns off impurities. Cast iron, whether braze welded or fusion welded, will require a flux. There are more impurities in the base metal to start with and the melting temperature is some 600 degrees lower, so that the impurities are seldom burned off completely. (By the way, braze welding is in almost every case the method of repairing cast iron preferred by professional welders, for a variety of reasons we will cover later.)

Fluxes almost invariably give off fumes that are at least noxious, if not poisonous, so that proper ventilation or respiratory equipment needs to be used.

FIRE HAZARDS

Given the extreme heat produced by even the smallest propane soldering torch, along with the open flame, the need for precautions against fire while brazing or braze welding is readily apparent. But precautions alone are not enough. After all combustibles and flammables are removed from the work area, you will want to make sure that any fire accidentally started can be quickly and safely put out. To do that, you need to either cut the temperature enough to eliminate combustion, remove the fuel supply, or cut the supply of oxygen to the fire. Since, in most cases, removing the fuel from the area will be difficult or impossible (it is extremely difficult to move a segment of a burning house wall), smothering or cooling below burning temperature are the only resolutions left to us.

There are four classes of fires of interest to the home welder. The

classes are important because they determine the type of equipment needed to cut down the flames and have them stay cut down. In some cases,
the importance is even greater, for a flammable liquid hit with a spray
of water can spread flames over a larger area determined only by the
force of the water spray.

Class A: Wood, paper, cloth, etc. Water is the best method of fight-
ing this class of fire, but is often not safe because of involvement with
electrical equipment or flammable liquids.

Class B: Flammable liquids. In such fires, smothering is the most ef-
fective treatment, as the vapor atop the pool or container of liquid is
what is burning. Cut off the oxygen to the liquid and it cannot form the
volatile vapor. Under *no* circumstances should water ever be used against
a class B fire.

Class C: Fire involving live electrical gear. Electrically nonconduc-
tive materials must be used to extinguish this class of fire. Again, water
should never be used, this time because of the hazard of electrical shock.

Class D: Burning metals. This type of fire is probably the least
known of the major classes. Metals of many kinds will burn at relatively low temperatures, magnesium being the best-known example since it
is used in flares and other military fireworks.

Fire extinguishers are classified according to their ability to handle
the different kinds of fires. Generally, in a home shop an extinguisher
capable of handling the three top classes, A, B, and C, will suffice, though
if you are working with magnesium you may wish to find and buy one
with a rating for class D fires. The capacity of the extinguisher should
also be expressed alongside each type of fire it is to be used on. For
example, most dry chemical fire extinguishers will do a fine job on class
B, C, and D fires, but will have a much lower capacity for use with class
A fires. The reason is simple: the dry chemical cannot penetrate the
burning material as could, say, water, to cool down the entire burning
section.

A good multipurpose dry chemical extinguisher should be the unit of
choice for most home shops, with, if you wish, a water extinguisher of
one type or another as a back up. The dry chemical extinguishers offer
several advantages other than being a good choice for almost all types of

fire. First, they don't freeze, no matter how cold the weather. They are most often made in a form that allows any of us to easily check the charge. And refills are readily available in almost any town of moderate size.

All of this emphasis on dangers and hazards and their prevention may well turn off anyone who is a bit timid. This is not my purpose. It is simply necessary to recognize the fact that any sort of welding equipment can prove an extreme hazard if improperly operated and maintained. This starts at the initial set-up of the equipment and goes right on through shutting the torch down and storing it. Follow the safety precautions and you should have few, if any, problems. Don't follow them, and you will join a statistical list that is a pleasure to no one. The idea of working at home with any kind of tool is to save time, money, and aggravation. Add to the expenditure of any of these, and there's no point to doing it yourself.

Chapter 2

EQUIPMENT SELECTION

Selecting brazing and braze-welding equipment can be more than slightly complex. Three fuel gases are in common use, and different types of oxygen producers and containers can be found at a quick look around any welding supply or large hardware store. The best spot to start is an examination of the work you plan to do, now and in the future. Because price variations can be so great, the savings in cost from the lowest priced single-gas torch to a full set of oxyacetylene torches, gauges, and tanks is on the order of $250, often more. (Fig. 2-1.)

Figure 2-1. The variety of torches available is immense. These are only a part of the line offered by a single manufacturer.

At the bottom of the brazing line are the gas/air propane torches. Almost all of these torches will provide enough heat to do small silver soldering (silver brazing) jobs, but they are totally inadequate to any kind of brazing involving the heating of large expanses of metal. Still, because of their low cost, and the low cost of the fuel gas, it pays for just about every homeowner to have some sort of propane torch on hand. Whether the job is silver soldering a clasp on a belt buckle or sweat soldering a pipe, these torches will do a fine job, at absolutely minimal cost. Most now cost under $10, complete with at least two torch tips and a cylinder of propane. (Fig. 2-2.)

Adding flexible hoses to a propane torch, and a larger array of tips, increases the versatility, but tends to come close to doubling the cost. The type of jobs that can be done is not really increased, but the ease of doing most of those jobs is greater.

From this point, you may wish to move to MAPP/air torches (in all cases, these fuel gas/air torches are also known as single-gas torches since pure oxygen isn't used to increase the flame temperature). A few months ago, there were only a half dozen of these on the market. Now that number has at least trebled and may well do so again. (Fig. 2-3.)

To my mind, MAPP is the gas of choice for a single-gas torch in al-

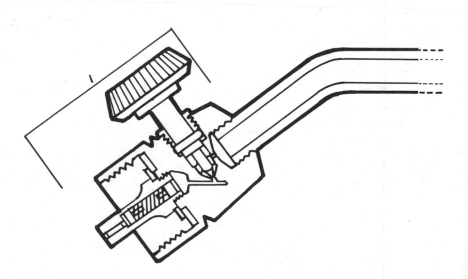

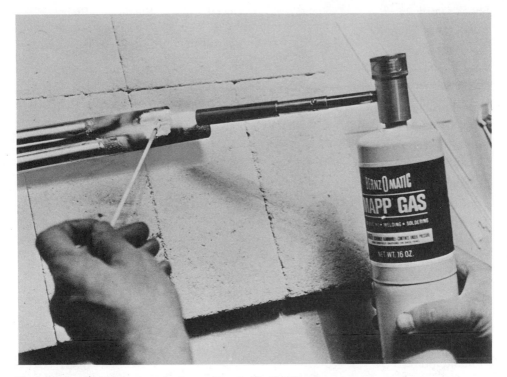

Figure 2-3. The Bernzomatic Super Torch for MAPP gas. BERNZOMATIC CORPORATION

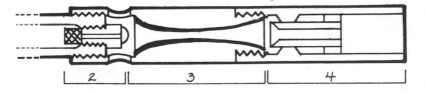

Figure 2-2. Schematic diagram of a torch. 1. Pressure regulating body. 2. Captive orifice/filter system. 3. Venturi intake system. 4. Flame attenuator.

	MAPP Gas	Acetylene	Natural Gas	Propane
SAFETY				
Shock sensitivity	Stable	Unstable	Stable	Stable
Explosive limits in oxygen, %	2.5–60	3.0–93	5.0–59	2.4–57
Explosive limits in air, %	3.4–10.8	2.5–80	5.3–14	2.3–9.5
Maximum allowable regulator pressure, psi	Cylinder	15	Line	Cylinder
Burning velocity in oxygen, ft/sec	15.4	22.7	15.2	12.2
Tendency to backfire	Slight	Considerable	Slight	Slight
Toxicity	Low	Low	Low	Low
Reactions with common materials	Avoid alloys with more than 67% copper	Avoid alloys with more than 67% copper	Few restrictions	Few restrictions
PHYSICAL PROPERTIES				
Specific gravity of liquid (60/60°F.)	0.576	–	–	0.507
Pounds per gallon liquid at 60°F.	4.80	–	–	4.28
Cubic feet per pound of gas at 60°F.	8.85	14.6	23.6	8.66
Specific gravity of gas (air = 1) at 60°F.	1.48	0.906	0.62	1.52
Vapor pressure at 70°F., psig	94	–	–	120
Boiling range, °F. 760 mm. Hg	−36 to −4	−84	−161	−50
Flame temperature in oxygen, °F.	5,301	5,589	4,600	4,579
Latent heat of vaporization at 25°C., BTU/lb	227	–	–	184
Total heating value (after vaporization) BTU/lb	21,100	21,500	23,900	21,800

Table 2-1. MAPP Gas Properties. AIRCO

Fuel	Flame Temp. (°F.)	Primary Flame (BTU/cu.ft.)	Secondary Flame (BTU/cu.ft.)	Total Heat (BTU/cu.ft.)
MAPP Gas	5301	571	1889	2406
Acetylene	5589	507	963	1470
Propane	4579	255	2243	2498
Natural Gas	4600	11	989	1000

Table 2-2. Heating Values of Fuel Gases. AIRCO

most every case. (See Table 2-1.) In air, you get a flame temperature several hundred degrees hotter than propane, and the tip design adds to the flame characteristics that make MAPP the better choice. In addition to the greater temperature, the BTU (British thermal unit) production of the primary flame is much greater for MAPP than for propane (almost double). This means the heat is delivered to the workpiece much more rapidly, thereby cutting down on the amount of fuel needed to provide a particular heat level. Acetylene has a high initial heat output also, though lower than that of MAPP, but propane is low, with a very high secondary flame heat transfer. MAPP offers a secondary flame heat level nearly as high as propane and far higher than acetylene (again, about double). Thus propane takes longer to transfer its heat, while acetylene transfers heat quite rapidly, but in a very small area right at the tip of the flame. MAPP gas transfers heat extremely rapidly, yet maintains a high heat level throughout the flame length. (See Table 2-2.) What all this means to the home brazer is quite simple: rapid heat-up of the workpiece at lower fuel cost, even though the price of the fuel gas is higher, and a less critical flame-to-workpiece "coupling" distance, which makes it easier to maintain the correct working temperature.

For those of us who already own oxyacetylene equipment, MAPP can still provide benefits, though some adaptations are needed because MAPP gas does not flow at the same rate as does the less dense acetylene, causing an imbalance in the oxygen flow (not enough oxygen can flow to get the correct type of flame). Some oxyacetylene torch heads can be converted by the owner by running into the torch tip 1/16 of an inch a bore tip drill three or four drill number sizes larger than the center oxygen

Tip Drill Size	Counterbore Drill	Tip Drill Size	Counterbore Drill
76	52	52	36
74	51	50	34
72	50	48	33
70	49	46	32
68	48	44	31
66	47	42	30
64	46	40	29
62	45	38	28
60	44	36	27
58	43	34	26
56	42	32	25
54	39	30	24

NOTE: All counterboring is to 1/16 in. depth. Different counterboring schedules are used for flame hardening heads due to special flame requirements.

Table 2-3.   Counterboring Welding Tips for MAPP Gas.   AIRCO

Drill Size of Tip	Inner Flame Length, in.	Regulator Pressure MAPP Gas	Range* Oxygen	MAPP Gas Consumption CFH	Metal Thick., in.
72-70	1/4	1-2	5-6	1-3	Up to 1/32
65-60	7/16	1-3	5-6	2-4	1/32-1/16
56-54	5/8	1-5	6-8	3-8	1/16-1/8
49-48	1	2-8	8-10	5-18	1/8-3/16
43-40	1-1/8	3-9	10-12	6-30	3/16-1/4
36	1-1/4	5-10	12-15	6-35	1/4-3/8

*For injector type equipment use 1-2 psig MAPP Gas and 25-30 psig oxygen for all size tips.

Table 2-4.   Substituting Acetylene Tips for MAPP Gas Use.   AIRCO

orifice. Others can be converted by contacting your nearest MAPP gas distributor (this gas is usually handled by Airco Welding Company distributors) and checking their catalog to see which brand and model torches can now be converted. (See Tables 2-3 and 2-4.)

And this brings us from the single-gas torches, ranging in price per kit up to about $30, to the more expensive oxy-fuel gas torches, with prices ranging from about $30 on up past $300.

Earlier, we mentioned the Bernzomatic model and the Solidox candle style models. Both of these inexpensive torches have advantages and disadvantages. Probably the biggest advantage is the price, especially if you keep a lookout for sales. The Solidox has a torch-style head on short hoses (about four feet), and uses a candle to produce oxygen. The candle is lit, by striking on a matchbook cover, and inserted in a tube, which is then sealed. In a minute or so, you'll have enough oxygen flowing to get a flame suitable for light braze welding. The limitations on braze welding are the burning time of the candle and the length of the hoses. The long-term candles last only about twelve minutes (and until you become familiar with the unit, it may take three minutes of fiddling to get the correct flame adjustment). The length of the hoses may limit reach when there is more than one spot to be brazed. The Bernzomatic uses throwaway oxygen containers pressurized to about 600 psi. The oxygen cylinder and the propane or MAPP gas cylinder are screwed into the torch head (standard right-hand threads for the oxygen cylinder and left-hand for the fuel gas), and adjusted by means of knobs at the top of the torch. Working time for brazing is about half again that of the Solidox torch, and there is a small gauge to inform you of the oxygen remaining in the cylinder. The limitation here is, once more, the working time per cylinder of oxygen. And the weight of two cylinders held in one hand, as is often necessary, makes the torch a bit unwieldy for really fine work.

Moving closer to professional equipment, we find units such as Airco's Tote-Weld. Weighing under thirty pounds, this outfit is similar to a few others. Because I own and use the Tote-Weld, my description will apply mostly to that brand. I opted for MAPP gas as a fuel gas, though the outfit can be purchased with tips and tanks for acetylene. Hoses are some eight feet long, and the torch resembles a smaller version of a full-

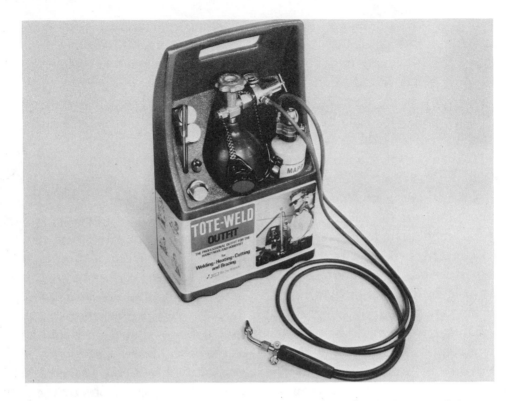

Figure 2-4. The Tote-Weld is an excellent compromise torch for the home workshop.

size welding torch, though it is much lighter and easier to handle. The oxygen cylinder is a twenty-cubic-foot model, and MAPP gas is supplied in a one-pound throwaway cylinder (this can be replaced with a 7½-pound refillable unit if you wish). (Fig. 2-4.)

This sort of unit will vary in retail price, but it is seldom found in complete form for under $160. Included with my Tote-Weld were cylinders, fuel, the torch and several welding/brazing tips, a cutting head attachment (the cutting capacity of this tool is about one inch), spark-lighter, goggles (of a type not suitable for those who wear glasses, unfortunately), a selection of brazing and welding rods, two bottles of flux, and a length of silver solder. About the only extra one needs buy to start right out is a pair of welding gloves.

The Sears Porta-Torch is similar to the Tote-Weld, except that it lacks an oxygen gauge on its regulator, which is more of a convenience than a total necessity. It's nice to know when you're running low, but most of us can keep pretty good track of usage anyway once we g t a bit of experience. The Porta-Torch, though, doesn't offer as great a cutting capacity (a half inch instead of a full inch), and because it is delivered by mail, it comes with an empty oxygen cylinder.

The disadvantage of the above units is their price, of course. Otherwise, it is all on the good side for the home shop. The units are reasonable in weight (the Sears unit weighs about two-thirds what the Tote-Weld weighs), have hoses long enough to allow a great deal of movement, have torches capable of precision adjustment, provide temperatures sufficient to braze-weld almost any material up to a quarter inch thick, and a few materials a bit thicker (it's always possible to work from two sides — where access is available — and double those thicknesses), come with cutting tips that will get you right on through half-inch material, and have a selection of tips, both cutting and welding, that will offer greater control over the work. In addition, you can, for hard soldering, simply cut off the flow of oxygen and use only the MAPP gas, which will provide more than enough heat for rather large jobs of that type.

Added to all this is the light weight and general ease of control of the torch heads on these tools. While a professional welder may sneer at this, such design considerations make matters a great deal easier for the novice, while providing the expert near pinpoint control of fine work, as for jewelry and decorative scrollwork.

A final step up is available to those having really heavy braze welding to do or those wishing to progress into the more complex field of fusion welding — oxy-fuel gas torches of full size. Complexity increases with these outfits, as does the price. It now becomes a simple matter to flatten the wallet, so much so that many, many people turn away from heat-application metalworking at this point. Basically, two types of full-size fuel gas/oxygen welding outfits exist, with great price variations in both ranges. The least expensive outfits are those with single-stage regulators; prices go up quickly for welding kits with two-stage regulators. The basic tanks and hoses are either the same or very similar, so that the increase in price is often almost totally due to the cost of the more precise gauges and torch. (Figs. 2-5 and 2-6.) In almost any home application, the single-stage regulators will serve admirably (as they will in all

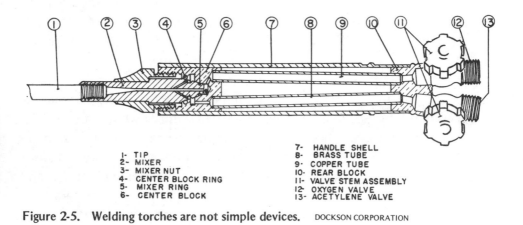

1- TIP
2- MIXER
3- MIXER NUT
4- CENTER BLOCK RING
5- MIXER RING
6- CENTER BLOCK

7- HANDLE SHELL
8- BRASS TUBE
9- COPPER TUBE
10- REAR BLOCK
11- VALVE STEM ASSEMBLY
12- OXYGEN VALVE
13- ACETYLENE VALVE

Figure 2-5. Welding torches are not simple devices. DOCKSON CORPORATION

but the most precise commercial use). My own oxyacetylene outfit is a
Harris Cost Cutter kit with single-stage regulators and a welding torch
that will do any brazing, braze welding, or fusion welding job I expect to
run up against in the future. Welding to a medium level is possible: this
can be classified as either braze or fusion welding to a half-inch thick from
a single side. Control of gas flow is precise enough to satisfy me, though
I should like to have my own two-stage set of regulators one day (at something on the order of sixty or seventy bucks apiece, they will wait for a
time).

Oxy-fuel gas welding outfits of the above kind are not supplied with cylinders. My suggestion would be that you either lease or purchase small
tanks, such as a twenty-cubic-foot oxygen tank, and an MC style (ten-cubic-
foot) acetylene (or MAPP) cylinder. The MC cylinders usually require an
adapter to fit the regulators, but the adapter costs under seven dollars.
These small tanks add to the portability of the outfit, while carrying plen-
ty of capacity for several hours' working time. Purchase can be expensive:
the small cylinders will sell for about $150, empty. Generally, the occa-
sional user will find it a better bet to locate a welding dealer in the area,
lease full cylinders, and return them inside of thirty days. Most dealers
will then charge only for the gas or oxygen used, about six to eight dollars

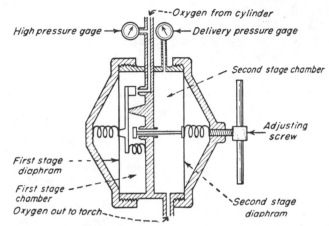

Figure 2-6.   Regulators are precision instruments and must be treated as such.   AIRCO

per cylinder. This also saves long-term storage of compressed gases and all the attendant problems.

Look for price ranges of single-stage welding outfits to start at about $110, with two-stage units beginning at $50 more and going upwards rapidly. Select an outfit with high quality neoprene hoses (standard length is twenty feet) covered with braided nylon. For our purposes, the best style of torch is the equal pressure model (also known as a medium pressure, or balanced pressure torch). This type of torch allows quick and easy adjustment of the flame, while being less subject to flashback than the injector type.

Flashback is a problem with acetylene torches, and for this reason you should give consideration to buying two inexpensive items known as check valves. (Fig. 2-7.) These check valves are inserted in the lines either at the base of the torch (preferable) or at the cylinder fittings. These rather simple spring-controlled devices prevent the backflow of gases, which can be a

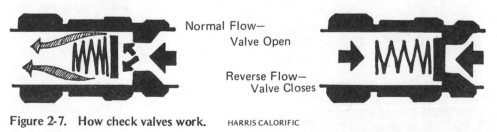

Figure 2-7.   How check valves work.   HARRIS CALORIFIC

ASSEMBLY INSTRUCTIONS

1. Be sure fuel valve is turned completely off. Turn fuel valve clockwise until snug.
2. Screw torch onto cylinder clockwise, by rotating tank.

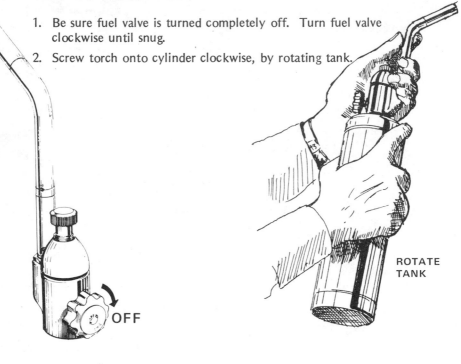

ROTATE
TANK

OFF

LIGHTING INSTRUCTIONS

1. Turn the fuel valve ½ turn counterclockwise.
2. Ignite torch with sparker or hold match at lower edge of burner tube.

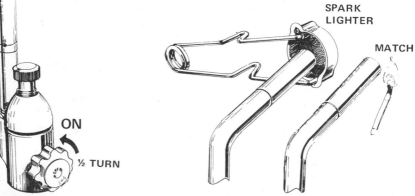

ON

½ TURN

SPARK
LIGHTER

MATCH

Figure 2-8. Assembly and lighting of single-gas torch. CLEANWELD

LIGHTING INSTRUCTIONS (Continued)

3. Turn fuel valve full on ¼ more turn counterclockwise.

4. Using the Adjusting Nut: Screw the nut down (clockwise) to increase the fuel pressure. (Recommended for High-Temp Fuels.) Screw the nut up (counterclockwise) to reduce the fuel pressure. (Recommended for propane and butane.)

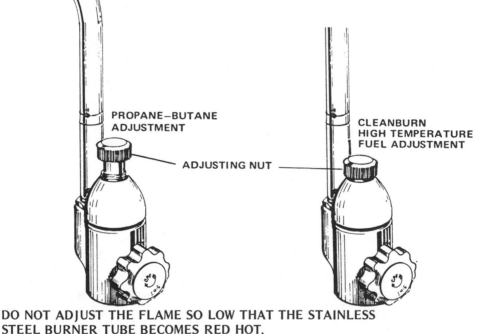

PROPANE–BUTANE ADJUSTMENT

ADJUSTING NUT

CLEANBURN HIGH TEMPERATURE FUEL ADJUSTMENT

DO NOT ADJUST THE FLAME SO LOW THAT THE STAINLESS STEEL BURNER TUBE BECOMES RED HOT.

5. The most efficient operation occurs when four distinct flame tips are seen on the large tip and when the inner blue cone is approximately one inch long on the fine point tip.

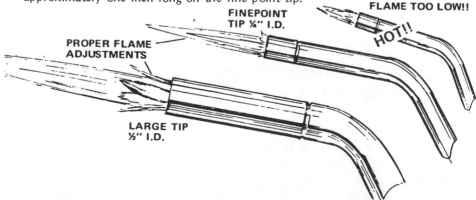

FLAME TOO LOW!!

FINEPOINT TIP ¼" I.D.

HOT!!

PROPER FLAME ADJUSTMENTS

LARGE TIP ½" I.D.

cause of severe problems when flashback or backfire occur. The check valves do not prevent either flashback or backfire, nor can they prevent the passage of flame, but they do reduce the number of equipment explosions caused by the recompression of gases in the hoses. (The heat of recompression can exceed $1800°$F.)

Once you have decided on the level of equipment you want to use, you've reached set-up time. With single-gas torches, this set-up is not extremely critical, so that simply following the manufacturer's directions will always provide good results. (There is so much variation in the adjustments, depth of mounting, and so forth, that to cover them all would require another book.) Simply make sure all threads are clean, dry, and unburred. Screw the torch into the propane or MAPP cylinder, make the adjustments needed, and light, using a sparklighter. (Fig. 2-8.)

More critical, and more standardized, are the procedures for the first set-up and general lighting of oxy-fuel gas torches (Fig. 2-9). Your first job is to secure the tanks. Either place them in a tank cart, or, as with the Tote-Weld, in their special holder. If this isn't possible, the tanks should be securely chained in place before work starts. Acetylene tanks must *always* be in an upright position. Remove the valve protection caps. Check the work area for open flame, stand to one side of the outlets and slowly crack the valve of the acetylene tank in case the threads of the outlet have been fouled with dust or grit. Next, check the oxygen cylinder for the presence of any dirt, grit, oil, or grease on the threads. Return the cylinder to the dealer if oil or grease are present in *any* amount. If not, crack that cylinder valve, again making sure to stand to one side and keep the valve away from any open flame. By cracking the valve, I mean just that: you want only a very, very slight opening of the valve in both cases.

Using the wrench supplied with the outfit, attach the oxygen regulator to the cylinder valve. Once the regulator is in place, you will screw the regulator adjusting handle counter-clockwise to release the mechanism and keep gas from entering the hose when it is attached. Two turns are usually sufficient for this.

The acetylene cylinder will be supplied with a T-handle wrench. The first chore here is to wire or otherwise attach that wrench to the cylinder.

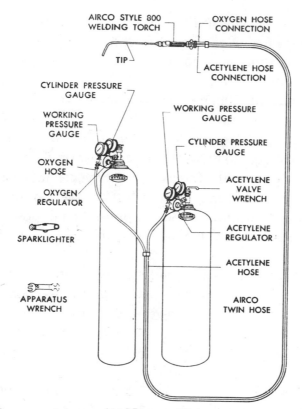

AIRCO STYLE 800
WELDING TORCH

OXYGEN HOSE
CONNECTION

TIP

ACETYLENE HOSE
CONNECTION

CYLINDER PRESSURE
GAUGE

WORKING
PRESSURE
GAUGE

WORKING PRESSURE
GAUGE

CYLINDER PRESSURE
GAUGE

OXYGEN
HOSE

OXYGEN
REGULATOR

ACETYLENE
VALVE
WRENCH

SPARKLIGHTER

ACETYLENE
REGULATOR

ACETYLENE
HOSE

APPARATUS
WRENCH

AIRCO
TWIN HOSE

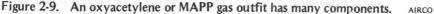

Figure 2-9. An oxyacetylene or MAPP gas outfit has many components. AIRCO

Now, attach the acetylene regulator to the valve and release its adjusting handle about the same two turns.

New hoses will have some talcum powder in their interiors. To keep from clogging the torch, it is necessary to remove this by blowing. The oxygen hose is blown out using oxygen or compressed air, if available. Then hold the end of the acetylene hose against the oxygen fitting at the regulator and allow about five psi to flow through it. Do not attach either hose at this stage.

Once this is done, blow out the acetylene hose by mouth to clear the oxygen. If compressed air is used, the job is safer, but with proper precautions both hoses can be purged with oxygen. Just make sure that both ends of the hose are free of oil and grease, that no oil, grease, or greasy rags are near, and that no other flammable material is around. Do not use acetylene to purge either hose.

Once the clearing of talcum powder is complete, you can attach one end of the hoses to their respective regulator fittings. The green oxygen hose goes to the oxygen regulator (which is usually trimmed in green, too), while the red fuel gas hose goes to the acetylene cylinder regulator (again, this will often be trimmed in red). The oxygen fittings cannot be attached incorrectly, as they are right-hand threads while the fuel gas threads are left-hand. To further aid in identification and prevention of possible tragic mix-ups, the fuel gas fittings will have nuts with a groove cut around them, while the oxygen fitting nuts are smooth surfaced.

The torch is next to be attached, and the fittings are threaded and marked in the same way as are the regulator fittings. As soon as the torch is attached, check to make sure its valves are cut off.

Now comes the time to open the cylinder valves. Oxygen is first. The valve is cracked and then slowly, very slowly turned until the gauge reads its maximum pressure. The best procedure is to simply crack the valve, then turn it only so far as needed to get the gauge needle to move slowly up the dial. Once the dial needle stops moving, you can slowly open the valve all the way. With the high pressure in oxygen cylinders, completely opening the valve right off will often rupture the gauge diaphragm, necessitating repair or replacement, both of which are expensive.

Next comes the acetylene cylinder, with its special valve wrench. Start in the same manner as with the oxygen valve, opening it only enough to cause the dial needle to move up scale. Once it reaches maximum pressure, turn the acetylene cylinder valve handle *no more* than one turn. That valve is never opened more than a single turn.

The gauges are now pressurized, while the hose and torch pressure remains nil. You have set the pressure regulators on the gauges at the off position earlier. Now, with the torch valves closed, turn the pressure regulator adjusters clockwise, or in, to set the hose pressure. At no time should acetylene hose pressure exceed fifteen psi!

With hoses pressurized, take your soapy mixture and test all fittings for leaks. Retighten if necessary, and then recheck for leaks. Do not over-tighten or force fittings.

Now is the time to purge the hoses in preparation for lighting the torch. First, open the acetylene torch valve, making sure that hose pressure is never more than fifteen psi. Use the regulator to up the oxygen pressure to about thirty psi, close the acetylene valve, and open the oxygen valve. (When using MAPP gas as a fuel gas, line pressures will be higher, but the purge is still needed.)

Torch lighting procedures vary slightly from manufacturer to manufacturer. In fact, sometimes the variation is not slight. One maker suggests opening the acetylene valve only a slight amount, using the sparklighter to light it off, and then mixing in oxygen. Another recommends cracking the oxygen valve, opening the acetylene valve fully and lighting off. Of course, both require a fair amount of adjustment to get the correct flame style afterwards. A pure acetylene flame will burn with a lot of black smoke, a yellow flame, and low heat. Welders vary in their reaction to instructions after they gain experience, so that what a person starts with is what that person is likely to stay with, no matter the torch or the instructions supplied with it. Originally, I began by opening the fuel gas valve and leaving the oxygen valve closed. A friend of mine began by cracking the oxygen and opening the acetylene valve all the way. Both ways have worked, for each of us, with no problems and no apparent danger through a wide variety of equipment, but it is still best to follow the procedure recommended by the company that designed and made the torch.

There are three flame types — neutral, carburizing, and oxidizing. Each is of value to the home braze welder at different times. The most basic and useful of all welding flames is the neutral flame; the carburizing, or excess acetylene, flame is often used in braze welding. Occasionally, very occasionally, there will be a use for an oxidizing flame when you're working on some metal other than iron or steel. There are major differences in flame appearance, extending even to the gas used. In fact, you can easily end up using an oxidizing MAPP flame because it has an appearance almost identical to that of the neutral acetylene flame. We will take another look at MAPP flames later.

Once the torch is lit, you will have, no matter the method of lighting the torch, an excess acetylene, or carburizing, flame. For certain types of brazing rod, this flame is recommended, in varying degrees. The carburizing flame is identified by the length of the feather that reaches beyond the inner cone. Carburizing flames are classified by the length of that feather: 1X and the feather is the same length as the cone; 3X and it is three times as long; etc. For most welding purposes, adjustment will continue until a neutral flame is achieved. No feather shows past the intense light of the inner cone, and the flame neither takes away nor adds anything to the metal of the weld (carburizing, or excess acetylene, flames add carbon, whereas oxidizing flames add oxygen). Extensive use is made of this flame in braze welding, and you should learn to identify it as soon as possible. The oxidizing flame is useful to us at only one time, actually, and that is when welding galvanized iron. The cone is exceptionally short, and the flame makes a harsh, hissing sound, much like a very angry, very large snake.

MAPP-oxygen flames differ in a few ways that are important, and since I've recommended this type of fuel gas as the best for home work, it is essential that we see just what those ways are.

The three basic flames are called by the same names, but in two cases the variations in appearance can ruin a braze weld for the beginning MAPP user (and this includes many, many experienced welders: in fact, the more experienced welder is likely to have a bigger problem since he (or she) will be used to the appearance of acetylene flames and will try and match those with MAPP gas). (Figs. 2-10, 2-11.)

CARBURIZING FLAME

NEUTRAL FLAME

OXIDIZING FLAME

Figure 2-10. MAPP flames.   AIRCO

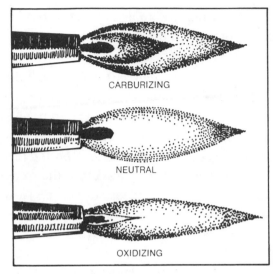

Figure 2-11. Acetylene flames.

MAPP-oxygen carburizing flames are very similar in appearance, and are just as useful, as are acetylene-oxygen carburizing flames for braze welding easily oxidized metals such as aluminum.

The neutral MAPP flame has a deep blue center cone, with no carburizing feather: the feather gradually disappears. You must keep an eye on this flame, as it will fairly rapidly change into the MAPP oxidizing flame with a light, intense blue inner cone of much greater length than is found on acetylene flames. This intense blue flame is almost identical in appearance to the neutral acetylene flame, and is useful for the same jobs: braze welding copper and copper alloys. It is the worst possible flame to use for braze welding ferrous (iron) alloys of any kind.

In any case, if you remember that the primary flame cone of the neutral MAPP flame is from 1½ to 2 times the length of the primary cone of an acetylene flame, you should be in good shape.

Once you have set the equipment up, gotten everything ready, lit the torch, and learned the appearance of the different flame types, you can begin considering the job of shutting down your gear. Shut down in the proper sequence is just as important as any other safety feature with gas welding gear. There's nothing really complex about it, but improper shutdown can cause problems from fire in the hoses to burst regulator diaphragms (or prematurely worn out diaphragms).

Once the flame is no longer needed, cut off the acetylene valve at the torch. This will suffice for shutdowns not to exceed five or ten minutes. For longer shutdowns, though, whether it is for an hour or two to run errands or for a weekend, you need to make sure the cylinder valves are closed and the hoses purged to release pressure on the regulators.

With the torch shut down as above, first close the oxygen cylinder valve and open the oxygen torch valve to release pressure. Now turn the oxygen regulator pressure adjusting screw out (counter-clockwise usually) to release pressure on the diaphragm. Close the torch oxygen valve. Shut down the fuel gas cylinder valve, and open the torch valve. Turn out the pressure adjusting screw on the fuel gas regulator and shut down the fuel gas valve on the torch.

Don't release both torch valves at the same time. Make sure the torch tip is pointed away from your body and away from any substances which might cause a fire hazard, when releasing hose pressures.

To restart your welding, follow the procedures detailed earlier in this section.

BACKFIRE AND FLASHBACK

Though some commercial welders use the words "backfire" and "flashback" interchangeably, there is a definite difference, and an important one, for flashback is more dangerous than backfire. Backfire is identified by a loud pop or crack as the flame goes out. It is a bit of a pain in the tail brought on most times by touching the tip of the torch to the workpiece. Sometimes the cause is an obstruction in the tip orifice, and other times an overheating of the tip. Close the torch valves, check the tip for obstructions, and, if none are found, make sure (by waiting, not touching) the tip has cooled down. Relight, readjust, and go on with the work.

Flashback is easily differentiated from backfire as it has a characteristic shrill squeal instead of a pop. It can be considerably more dangerous, as it is the feeding back of the flame into the mixing chamber of the torch.

This can cause the torch to burn, the hoses to burst. In the case of flash-back, shut down the torch valves immediately, then go back and shut down the tank valves, if the noise and burning continue. The causes of flashback are kinked hoses, a loose tip, mixer seat (a part of the torch), incorrect pressures, a clogged torch tip or orifices, or failure to purge the hoses before lighting the torch.

In the case of flashback, the torch should not be relit until a thorough check of the causes of the problem has been made. Whatever the cause, it must be corrected before relighting or the problem will rapidly recur.

TROUBLE	PROBABLE CAUSE	REMEDY
Welding Tip popping	· Too close to work	· Move further from work area
Flames not clearly defined, smooth or even	· Dirty tip	· Clean with tip cleaner or replace tip
Cutting Tip popping	· Too loose · Nicked seat	· Tighten tip nut · Replace tip
Leak around needle valve	· Packing nut loose	· Snug packing nut
Difficult to light	· Needle valve open too wide	· Partly close needle valve
Flame change when cutting	· Oxygen or acetylene cylinder almost empty	· Replace cylinder with full one

Table 2-5. Troubleshooting Chart.   SEARS, ROEBUCK AND CO.

Chapter 3

BRAZE WELDING

Braze welding is no more complex than simple brazing (covered in the next chapter). Certain torch handling details add to the time needed to pick up the skills, but the joints are actually a bit simpler to set up, and the entire process is much less critical than fusion welding.

The brazing processes, including braze welding, require temperatures of $800°F$. or above. There is no clear break as to the point where brazing leaves the fold and braze welding jumps in, insofar as temperature of the materials being joined is concerned, for the primary difference between the two processes is in joint design, with a resulting difference in fluxes and filler metals, as well as in joint strength.

Braze welding uses the same joint designs as does fusion welding. (Fig. 3-1.) Filler metals consist of various alloys, including bronze or brass, often with silicons and other substances alloyed in to give different properties. In braze welding, the joints are *not* sealed by capillary attraction as they are in brazing. The base metal is heated to about a dull cherry red, but not to the point of actual base metal melting, and the filler rods are flowed into the joint to provide great strength and ductility. The filler metals, in a properly braze welded joint, bond so closely with the molecules of the base metal that the resulting bond is exceptionally strong. Cleaning by mechanical and chemical action is essential to the process, as the joint cannot be strong if tight contact between filler metal molecules and base metal molecules is interfered with by contaminants.

Because the base metal is not melted, braze welding is much simpler to handle than is fusion welding. The occasional welder doesn't have to worry so much about burnthrough and metal warping (though thin sheet metals must still be heated with care or warping will result). The speed of welding is also increased since the metals need not reach the greater temperatures required by fusion welding, so that you save time and gas. The effects of expansion and contraction are lessened, which allows simplified repairs to such metals as cast iron.

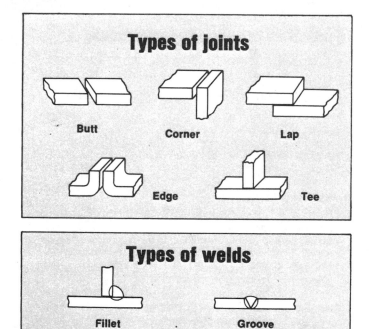

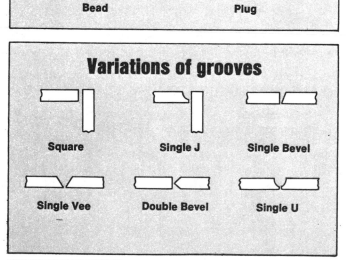

HOBART BROTHERS COMPANY

Figure 3-1.

On top of all this, the filler metal used has a high yield factor until the cooling temperature drops below 500° F. The advantage here is that the yield of the filler metal allows the locked-in heat stresses in the base metal to work themselves out, adding to the strength of the final repair. Without special treatment, it is often possible to get stronger welds using braze welding in materials such as cast iron than it is when using fusion welding (with special pre- and post-heating processes, the fusion welding job can be made stronger, but the time required, added to special equipment needed, make the small gain just about totally worthless from the point of view of the hobby or other occasional welder).

JOINT PREPARATION

No matter the metal being braze welded, joint preparation is of great importance, as it is in just about every metal jointure method. For metal under ¼" thick, the edges must be chipped down to bright metal. If the metal is over ¼" thick, you'll need to bevel each edge, using about a 60° to 90° vee. (Fig. 3-2.) While a cutting torch can be used to bevel the edges, the best practice for smaller joint lengths is to use a cold chisel. Using a torch or a grinder tends to spread the graphite particles in the base metal all over the edge of the joint, and this prevents the copper in the filler metal from doing as good a job of "tinning." Tinning is the forming of the molecular bond that is so important to joint strength. In metals other than cast iron free graphite is not present to an extent that need worry us, so that mild steel, for example, can be beveled in any convenient way. Should the cast iron under repair be so large that you cannot or do not wish to chisel a bevel by hand, then go ahead with the grinder or torch, but make sure you then use the torch, set for an oxidizing flame, to sear the edges well back from the joint. (Heat the joint to a dull red, and then let it cool before starting the braze weld.) You may find it necessary to use this process anyway if you are working, as you are most likely to be, on such things as engine manifolds, engine blocks, and other parts, frequently or always covered with oil, grease, and other grime, and then kept for hours at a time at a relatively high heat. In such cases, make sure the part is removed to a

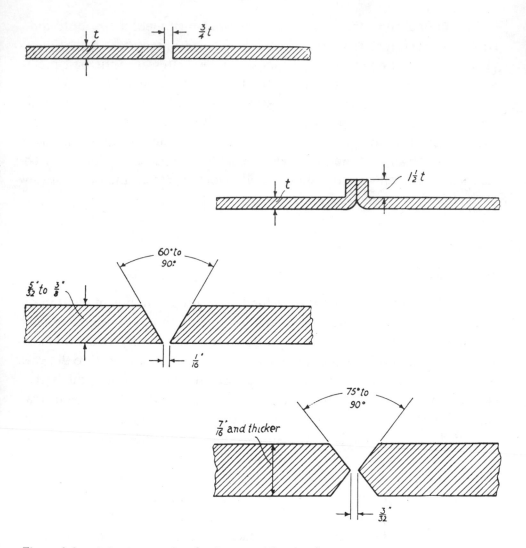

Figure 3-2. Joint preparation for braze welding is of great importance. AIRCO

place where other engine parts won't be damaged by the repeated application of heat needed to allow proper tinning.

Joint preparation continues on into joint design. Gas welding of any kind has a wide array of joint designs, each with a use in a specific application. In every case, the purpose is to allow the filler metal to penetrate to the root of the weld so that your joint will have the greatest possible strength. For metals up to about ¼", no special designs are really needed; the flowing filler metal should easily penetrate the joint. For thicker pieces of metal, the temperature loss from top to bottom may be extreme if the shape of the joint impedes the flow of hot metal. So bevels are used, in a variety of styles, to allow the filler metal to get where it needs to be at the temperature needed for correct bonding. Torch movement is also important to the flow of filler metal, and will be covered shortly.

A simple butt weld, done in a flat position, requires only that the ends of the pieces to be joined be clean. No grooving, beveling, or other preparation is needed. The distance between the metals being joined should be equal to one half the thickness of the base metal.

A single vee butt weld is just about as simple, except that the bevels are needed, as shown. A similar style of weld is formed with a corner weld where it is possible to leave a 90° gap between the sheets or plates of metal. With the single vee butt design, each plate is beveled at 45° to form a total of 90°, leaving a land for the root of the weld of about one-third or a bit less of the base metal thickness. This land will become smaller in relation to the thickness of the base metal as that base metal increases in thickness.

When a tee section is needed, either as a piece to be added to a project, or as a repair on an original part, a type of weld known as a fillet is used. Fillet welds are also exceptionally handy in making lap joints. For thinner metals, again no preparation other than clean-up and fluxing is needed. For metals over ¼", a single bevel of 45° will suffice for simple tee fillet welds.

One other type of weld is of importance in some operations, though the flanges that require edge welding, as shown, will not be found too often around the home. There is no great difficulty with this form of weld, but you will usually be working with relatively thin metals, and often soft metals (aluminum or copper sheet), so it is a good idea to clamp the works together.

For heavier pieces of metal, you will need to have a good-size root opening, but that will, even with a 90° bevel, often not provide a strong enough joint. Too much build-up of filler metal in a single spot can lead, even with braze welding on thicker sections, to burn through of the root section.

There are two ways to get around this, with the two-sided bevel being preferable where you have access to both sides of the part being repaired. When such access is not possible, you simply begin with a smaller portion of the beveled groove being filled in the first pass, getting a good, solid root for your joint, and then move back and make as many other passes as are needed to complete the weld (Fig. 3-3).

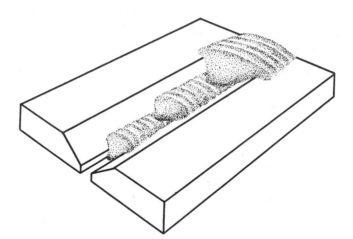

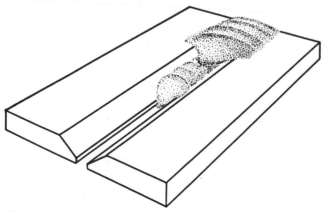

Figure 3-3. When the metal is thick, you will often need to make more than a single pass while laying down filler metal. AIRCO

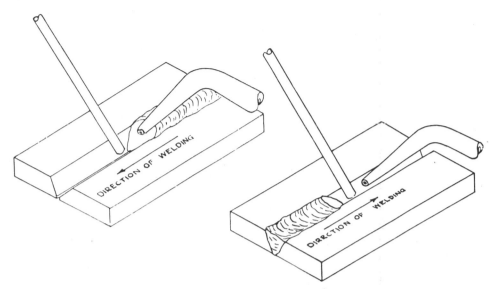

Figure 3-4. (Left) Forehand welding. Figure 3-5. (Right) Backhand welding.　AIRCO

TORCH HANDLING

In the last chapter, we covered getting the torch set up for use, and adjusting for the correct flame type, as required by the job. Sorry to say, that isn't quite all there is to the skills needed for obtaining great weld-joint strength.

First, there are two methods of handling the tip of the torch in relation to the rod of filler metal. As a start, the forehand or puddle method (Fig. 3-4) is used. With the forehand style, the welding rod is ahead of the torch tip as they move along the weld. The rod is arced, or circled, in one direction, while the torch tip moves through the same pattern but in the opposite direction. There are two disadvantages to this method for the hobby welder: first, because of varied movements of the tip and rod needed to keep the rod in the torch flame and the edges of the joint preheated, it is moderately difficult to learn. Second, those same oscillations force you to open up the bevels in a joint to make enough room for the movements. Simpler to learn is the backhand movement of the torch (Fig. 3-5). In the backhand method the torch tip moves ahead of the filler rod as the

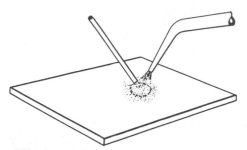

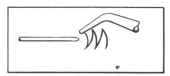

Figure 3-6. Torch and filler rod movement are important, too. Top: The backhand welding method requires you to move the filler rod. Bottom: The forehand method requires movement of the torch tip. AIRCO

weld is made. Backhand welding is carried out by moving the torch along the joint at the correct pace, while following it with the filler rod, which may be oscillated if you wish. The torch tip moves in a straight line, though. (Fig. 3-6.)

Though the backhand method is easier to use for the sometime welder, most professionals seem to prefer the forehand method for metals 3/8" thick or less. Once this thickness is passed, you'll get much better penetration and joint strength with the backhand style of welding with greater ease of control. In addition, you won't have to open up the vees in your bevels as much as you would with the forehand style. Where a forehand weld will require a 90° bevel (45° on each piece), you can cut that to 60° (30° on each piece) for a backhand butt weld.

Torch tip and filler rod angle are also of great importance when braze welding (Fig. 3-7). No matter the style of welding, the thicker the metal, the more nearly vertical the torch tip needs to be in order to concentrate the heat and get good bonding of filler metal and base metal. (Fig. 3-8.)

Backhand welding is almost always the preferred style for overhead welding. The reason is simplicity itself: molten metal, like all liquids, wants to react to gravity. Filler metal forms a puddle of molten metal, and the puddle is smaller with the backhand method, thus easier to control in vertical and overhead welds. In all cases, the narrower bevels and simpler process of backhand welding results in a quicker job, using less gas and fewer filler metal rods. Even when you are welding at home, cost is very important. Bear this in mind when you are learning a style of braze welding torch manipulation that you intend to be your main method.

TORCH ADJUSTMENTS FOR BRAZE WELDING

An oxyacetylene or oxy-MAPP gas torch would seem "ready to go" on most kinds of metal once it is set up for any kind of welding. This statement misses the truth by quite a bit. Braze welding requires different flame adjustments and different tip sizes for particular filler rod sizes and metal thicknesses than docs fusion welding, so that any instructions which may come with your torch covering fusion welding will need at least a bit of modification.

To start with, the filler rod will often specify the type of flame to be used. In many cases, a slight excess fuel gas flame will be needed for best flow results, while in others, most notably gray cast iron (the kind we're most likely to run into), a slight oxidizing flame works best. With the noted exceptions, though, you will still find yourself working most of the time with a neutral flame.

Torch tip selection is invariably a problem for the beginning welder. It can, in a few cases, be a problem for the expert, to the point where he may have to try one or two extras to get the desired results. It simply is not always possible to look at the job and say, "Well, I think a number 4 will do it this time." The factors affecting tip size choice are the thickness of the metal, the skill of the welder, the metals being worked upon, the position of the weld, and the desired speed of welding. Manufacturers' charts will cover tip sizes for use with particular metals up to or below certain thick-

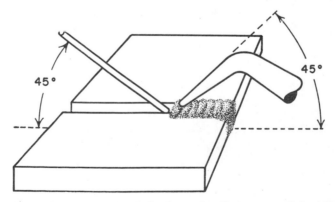

45° 45°

Figure. 3-7. Torch and filler rod angles help determine the success of the joint. AIRCO

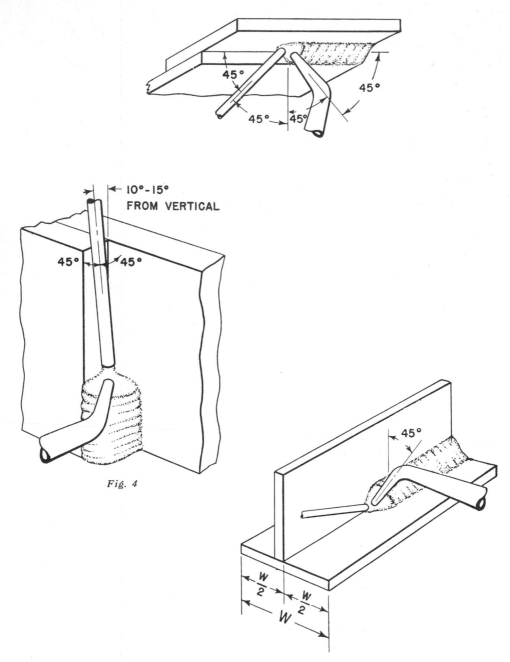

Figure 3-8. As the joint and weld type change, so do the angles at which you need to hold the filler rod and the torch. AIRCO

nesses. Assume a modest size tip, with a drill size for the orifice of about 60, and you would be prepared to fusion weld metal up to about 3/32" thick. The same thickness of metal being braze welded would require a tip with a drill size of at least 56, one size larger and usually classified as a tip for metals up to 1/8" thick. For 1/8" braze welding, you would need a tip offering a drill size of 53, and a fusion welding limit of 3/16". Since tip sizes are different, at least in the designations, from manufacturer to manufacturer, there is not much real point in supplying you with a tip size chart here and saying go to a number 171. That may be fine if you happen to have a torch from that maker, but otherwise you are still swinging from one horn of a dilemma and looking for a place to land. Do it the simple way. Go to the supplier who sold you your outfit and ask him to get you a chart, a Xerox of a chart, or some such, for the tips as supplied and designated by that manufacturer. Then go ahead and use the tip one size larger than is recommended for the fusion welding job. As an indication of the differences in tip sizes, no matter the numbering system, the Harris No. 1 tip is meant to be used with metals up to 1/32" thick, while other manufacturers will offer you a No. 1 tip meant for use with metals up to 3/32" thick. This has nothing to do with quality, but simply means one maker offers a different size No. 1 tip than does another. Possibly than any other.

So tip charts are simply a basis for starting to select the correct tip for the job. If all goes well, then the tip is the correct size. For braze welding, you need a tip one size larger than for fusion welding, as a start. And the backhand method requires a larger size tip than does the forehand. Certain metals will force the choice of a larger tip: the faster the metal conducts heat away from the weld area, the larger the tip size needed. Aluminum is a prime example of a metal that requires a larger-than-listed tip size. In general, the bigger the joint to be made (not length, but thickness), the larger the tip needed.

If all of this ends up with your starting with a tip, and finding that the gas pressures are giving a hard flame that blows molten metal away from the joint, or that the torch consistently backfires or pops out, then it is time to make certain adjustments in the size of the tip. Only experience will tell you exactly what movement to make when these troubles occur,

Figure 3-9. Braze welded joints are quickly made if enough heat is used and the joint has been correctly prepared.

as torch handling skills have a definite effect, as does getting the regulator pressure correct. (Figs. 3-9, 3-10.)

FLUXES

Once the torch is set up, the joint design decided on, and the metal basically prepared for braze welding, it is time to pick up the flux container and make sure you have got the correct chemicals. Because so much braze welding is done these days, the selection is rather simple. Three things need to be considered if the flux is to do its job: (1) the temperature of the work being done; (2) the method being used to make the joint; and

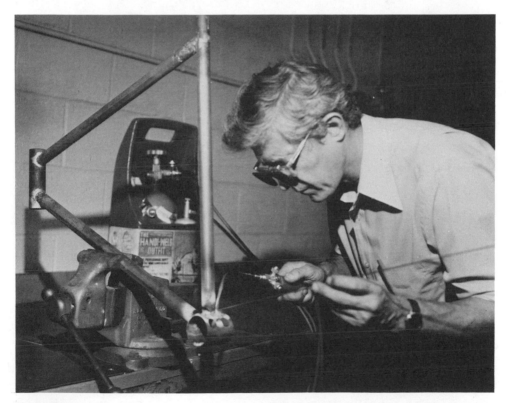

Figure 3-10. Oxy-MAPP torches such as this one provide the heat needed for larger joints or repairs.

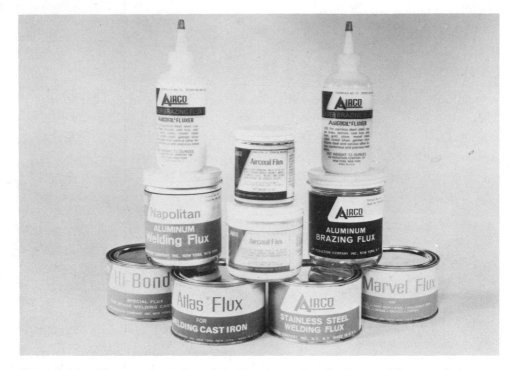

Figure 3-11. These are just a few of the fluxes you may find at a welding supply house.

(3) the types of metal being joined. Whatever flux you use will be compounded to handle slight variations in the above, but if you use a low temperature process (say, under 1100°F.) and a high temperature flux, you can bet that enough impurities will remain to interfere with the joint bonding, thus cutting joint strength badly. One of the purposes of flux is to lower the melting temperature of the oxides which will interfere with bonding, which then allows the impurities to float, as slag, to the surface. With cast iron and metals other than steel, many of the oxides have a melting point *higher* than that of the metal, so that chemicals are essential to a clean, strong joint. Since in braze welding the base metal doesn't melt anyway, we have even greater need of highly active fluxes. (Fig. 3-11.)

Again, because of the numbers of fluxes and filler rods on the market, it is impossible to tell you exactly what flux to use with what rod on what metals. All flux manufacturers and filler rod makers offer information on the varied applications of both materials; I recommend that you follow their instructions as closely as you can. A quick look at an excerpt from one company's cut-length welding rod catalog will show exactly what I mean.

Airco No. 1010 AWS A5.7
(Class R Cu Si-A) Silicon Bronze Rod
Description: No. 1010 silicon bronze bare gas welding rod contains approximately 2% silicon and 1% manganese and has a weld deposit tensile strength of 50,000 psi minimum.
Application: No. 1010 silicon bronze rod is recommended for all position welding of copper, copper-silicon, and copper-zinc base metals to themselves and also to plain and galvanized steel. It can be used for TIG, oxyacetylene, or carbon arc welding. Keep the weld pool small to assure rapid solidification and to avoid contraction strains. Use with Airco Marvel, Hi-Test, or Hi-Bond Flux.

This rod is only one of many that Airco offers for various braze welding purposes (Airco No. 27 is actually their general purpose braze welding rod), but the depth of instructions, flux recommendations, and so forth are just as good for the others. Several of the rods are available coated, as well as bare, so that there is even less real difficulty in selecting a flux. You don't have to bother when using a coated rod.

BRAZE WELDING PROCEDURE

Once joint preparation, torch set-up, and flux choice are ready, pick up the filler rod to be used — make the choice on the manufacturer's recommendation — and light and adjust your torch. The first thing to do is to check the work to see if it will need preheating. Most items of cast iron *must* be preheated to cut down on cracking and other problems caused by changes in internal stress during heating and cooling. Other metals may or may not need to be heated. Again, check the manufacturer's recommendations for the rod being used. Airco recommends preheating all materials braze welded with their Airco No. 20 AWS A5.7 Naval Bronze Rod to at least 1300°F., whereas other rods have different recommended temperatures. Set the flame. With one exception, Airco's braze welding rod requires a slight oxidizing flame for general use; others will need a neutral or carburizing flame.

The first step in a butt joint that isn't clamped should be a tack weld or two along its length. A tack weld is the simplest of fusion welds to make with the gas welding gear. Simply heat the base metals until they start to flow together, then remove the heat, and let cool. (Fig. 3-12.)

Now, with steels and cast iron, heat the metal until it glows a dull red, keeping the end of the filler metal rod in the flame so it will also preheat. Dip the rod in flux to check rod heat. If the flux sticks, preheat is just fine.

With the rod fluxed, the base metal at a dull red, and all else set, place the rod in the flame, using, preferably, the backhand method. Now comes the part that proves just how easy braze welding really is. If the base metal is too hot, the filler rod will melt very quickly and the filler metal will bubble and possibly even spit a bit as you apply the first, thin, tinning coat of filler metal. Tinning won't take place if the base metal is too hot. Back off a bit and let things cool down. If the base metal is not hot enough,

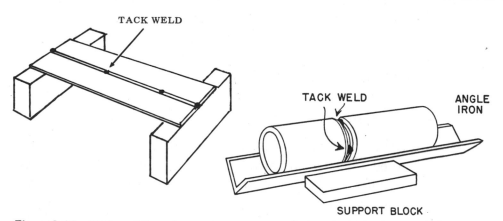

Figure 3-12. Tack welding. In some cases a fair amount of ingenuity is needed to keep the joints in place while braze welding takes place. AIRCO

tinning will also not take place, for the filler metal will simply form little balls on the joint. Apply more heat.

When the temperature is right, the flux is working, and you are moving at the correct pace, the filler metal will form a small puddle (using the back-hand welding method: if you use the forehand method, the puddle will about double in size) and flow along the joint easily without causing problems.

Use as many passes with filler metal as are needed to fill the joint being made. Don't move the parts until after the joint has cooled down quite thoroughly. And do not attempt to cool the joint by tossing water or oil on the part. In fact, retarding the cooling process with some metals, most particularly cast iron, is a good idea; it prevents the setting up of internal stresses caused by rapid cooling. The longer cast iron takes to cool, the greater will be the ductility of the joint and the surrounding base metal (ductility, by the way, is the ability of metal to be deformed without breaking). Clean the finished joint with a wire brush or chipping hammer and the job is done.

Chapter 4

BRAZING

As we've already said, brazing is a lower temperature method of joining metals than is braze welding; while brazing temperatures start at 800°F., they seldom reach as high as 1800°F., with most falling well below 1500°. Soft soldering stops at about 740°. Joint design is another major point of difference between braze welding and brazing. Welding joints cannot be used in brazing, for only a thin film of modestly strong filler metal is deposited in the brazed joint. Joint design is the result of another difference between the two methods. The filler metal moves into brazed joints by capillary attraction, a physical phenomenon that takes place when a solid and a liquid meet in a way that causes the surface of the liquid to form other than a flat shape. Filler metal being worked this way will flow uphill, sideways, or just about anywhere the correct heat is applied. The filler metal flows toward the heat.

It is at this point that the cheapest torches come into use. With care, even a small propane torch will braze, and single-gas MAPP torches will even do sizeable jobs. While the torches are relatively inexpensive compared to those used for braze welding, the filler metals are often extremely expensive. Much of the brazing that is done today is silver brazing, and a short length of that filler metal can be very costly. It is fortunate, then, that only small amounts are needed.

The uses of brazing are many. First, the joint, though not as strong as a braze-welded joint, is many times stronger than a soft-soldered joint. Next, it is, like braze welding, a process by which dissimilar metals can be easily joined. Distortion, because of the relatively low temperatures, is even less than with braze welding. Then, too, if you know where to look, you can find brazing alloys in many forms. If you wish to join pipes together, there are specially made washers that will fit in the joint, requiring then only that you heat the joint for the seal to be made. Sheets of filler metal can be clamped between pieces of metal to be joined and the pieces thus brazed together.

The advantages are many. The ease of application is great. The expense is low. (Silver brazing is not the only style. Various alloys that cost much less are also to be found: for such alloys, and for special shapes of filler metals, check with your local welding supplier. Very little of this stuff is carried in hardware stores, though large plumbing supply houses often have pipe washers and rings in a wide range of sizes and alloys.)

BRAZING JOINT DESIGN

While a brazing filler metal forms the same sort of molecular bond as does a braze-welded filler metal, the bond is not as strong, because the filler metal is not as strong. Because of this, the large, open joints of welding cannot be used. Separation in the joint should be from 0.002" to 0.006" when using silver or copper alloys for the brazing. When aluminum is brazed, the joints are opened up to 0.006" to 0.015". According to Airco, laboratory tests show that larger joint sizes decrease strength greatly, while, of course, smaller joint openings impede the capillary flow of the filler metal, again messing up the joint.

The drawings of Fig. 4-1 show the various designs for a number of different joints, along with possible bracing to increase strength. The chart (Fig. 4-2) shows how the joint clearance works to increase and decrease joint strength. With too great clearances, we can see that the joint strength rapidly drops off the rather weak "as cast" strength of the filler metal used, while fitting a joint right in the mid-range of the recommendations provides about double that holding power.

Surface preparation of the joint for brazing is of as great importance as for any other metal-joining method. First, mechanical cleaning is needed to take away the heaviest oxides, dirt, oil, and other impurities. This is done, depending on the size, shape, and style of the joint, with fine steel wool, emery cloth, and, sometimes, plain old soap and water. Then the joint is designed to provide the greatest support possible, so that strength doesn't suffer. Often this will simply mean providing more joint surface for the bonding action that takes place. In other words, if you are joining a section of strap metal, as you might be when repairing a bandsaw blade, you will make the joint on a 45° diagonal instead of straight across

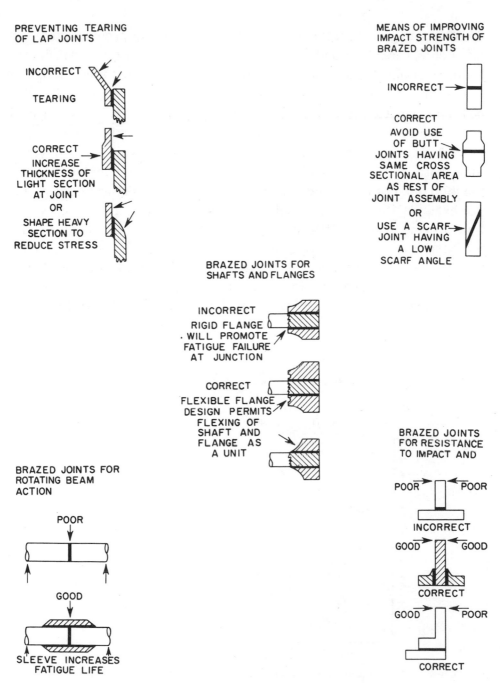

Figure 4-1.   Joint designs for brazing.   AIRCO

the strap. In general, it is best to use a lap joint, as this provides more surface contact area for bonding action than can a butt joint, but, as with the
bandsaw example, this is not always possible, so you must think things
over a bit and come up with a compromise. In almost every case, it will
be a simple matter of increasing the bonding surface by making a diagonal
joint (also called a scarf joint) instead of a straight-across butt joint.

When working with capillary attraction (as you do with brazing), you
should use the finest grits available in mechanical cleaning methods so that
the joint will not be opened beyond the strength limits already described.
For example, if you are silver-soldering copper tubing or pipe, use a very
fine grit emery cloth or fine steel wool to clean the ends of the pipe and
the interior of any elbows or other joints. Don't use rough grit sandpaper,
as that will not only cause deep gouges that may interfere with bonding,
but will also leave behind abrasive grit that will prevent capillary flow.

Once the mechanical cleaning is complete, you must select a flux, just

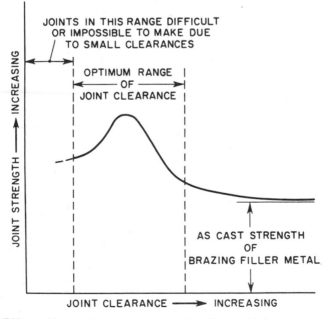

Figure 4-2. Effect of joint clearance on strength of brazed joints.   AIRCO

as you would when braze welding. Here, however, there are slightly different needs, though similar. The flux must provide a film of protection to prevent oxidation of the already cleaned joint surfaces, while dissolving any light oxides that may have been left by mechanical cleaning (or that have formed since the cleaning). Then the flux must also aid the filler metal in flowing easily into the areas of the joint where it is needed. This last action of brazing flux means you need to use care when coating the joint surfaces with flux, as the filler metal will flow onto fluxed areas even though they are outside the joint. While this is seldom a major problem if you're working with decorative material, such as jewelry, it is sloppy looking. And as the filler metal for this type of work tends to be quite expensive, the use of excess is a waste of money.

Again, the type of flux used will depend on the filler metal, as the varied compounds of filler metals will have different melting points. The flux must protect the joint from below the temperature range at which the filler metal starts to become liquid to a range above that point. Generally, the filler metal will come with recommendations for types of flux to be used.

Flux for brazing (silver brazing or soldering) serves one other important purpose, one that makes the job exceptionally simple, even for a beginner. At about the boiling point of water (212°F.) silver brazing fluxes will also boil, and the next visible change is a powdery look that appears several hundred degrees later. When the flux turns to a clear fluid that doesn't flow too easily, you will know you have passed the temperature range for soft soldering (800°F.). When the correct brazing temperature is reached, the flux will turn dark and flow easily, and you will know it is time to apply the filler metal to the joint, while keeping the heat at about the same level to maintain the temperature. With externally applied filler metal, as in sweat brazing of pipe or tubing, you need only look for a line of filler metal at the joint to find you have completed making the joint. Remove the heat and allow the joint to cool down well before moving anything.

For other types of joints, the indications will be present, though those made with external application of filler metals are easiest to determine, as the filler metal will melt upon application to the joint and flow quite distinctly into the joint. Where rings, plates, gaskets, or other specialized types of filler metal shapes are used, you may have to keep your eyes open to see the line of metal forming: it will be there. (Fig. 4-3.)

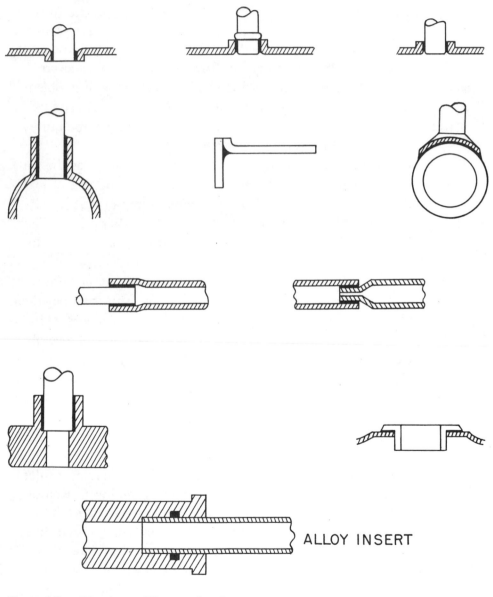

Figure 4-3. Alloy insert filler metal styles. AIRCO

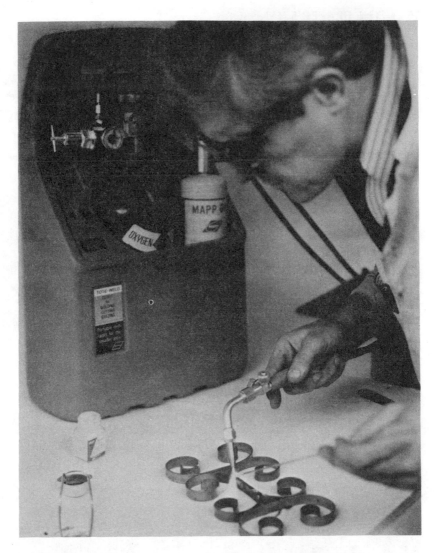

Figure 4-4.   Oxy-MAPP gas torches can also be used for brazing. Simply cut
off the oxygen and use only the fuel gas if the parts being joined are light.

HEAT APPLICATION IN BRAZING

Heat application in brazing differs markedly from heat application in braze
welding. To start, you will, in almost every instance, be using a single-gas
torch, such as the new Bernzomatic Jet Torch II. (This new style of propane torch has a swirl pattern flame that applies the heat more quickly to
the surfaces being joined, so that brazing is possible — for very light metals, braze welding is often possible with this torch, too — as it is with single-gas MAPP torches.) (Fig. 4-4.) Once the joint is prepared, including

mechanical cleaning, fluxing, and the assembly of the joint parts, the heat is applied. Keep an eye on the flux as it passes through its stages of use as a heat indicator. Once the flux becomes watery and dark, apply the filler metal to the joint, with the heat held on the side *away* from the filler metal. Heat, in brazing, is not applied directly to the filler metal. The torch

Figure 4-5. MAPP torches are ideal for hard soldering (brazing) pipe runs.

is moved as constantly as possible. If you are working with a pipe joint, start with the heat applied on one side, the filler metal on the other and move both around the joint until the line of filler metal is complete. Remove the heat and let cool. (Figs. 4-5, 4-6.)

If all is well, the filler metal will flow very freely, almost seeming to zip

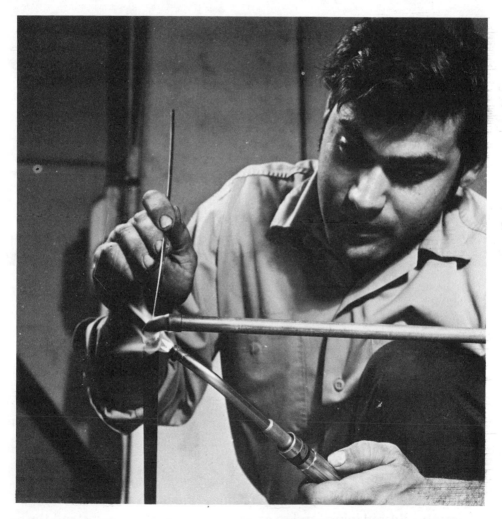

Figure 4-6. Capillary attraction means not having to move the work to get the filler metal to flow upwards. Note that the heat is applied on the side opposite the filler metal.

Figure 4-7. All sorts of decorative designs can be made, for just about any purpose, by brazing.

up into the joint: the filler metal will move toward the heat, so that you can forget about any need for taking a vertical pipe joint and placing it on a horizontal to get a complete joint seal. The molten filler metal will go anywhere you wish it to, as long as the area you desire has been fluxed and is heated. This, of course, is a great effort saver, and makes silver brazing an exceptionally handy way to seal joints for a great many things.

According to Airco, silver soldering or brazing is in heavy use in many areas, including automotive hydraulic lines (the joints are much stronger than soft soldered joints, and so are much better able to withstand the pressures built up, for example, in a brake system or a power steering system), refrigeration and air conditioning equipment, electrical joints that may receive more vibration or stress than normal, and in marine uses for various cable, hydraulic, and other fittings. And as we've already mentioned, it is widely used in making jewelry, whether at home or in a factory.

Once the joint is completed, your work is not quite over. Brazing fluxes tend to be highly active, with resultant probabilities of corrosion after a short time has passed. These fluxes, whether of the corrosive type or not, will also cause paint or most other finishing products to lift right off the surface, if it is one that requires such a finish. Generally, a good washing down with hot, soapy water will get rid of all excess flux. If the joint is to be decorative, you may wish to use a file, emery cloth, or other such tools to remove any excess metal. (Fig. 4-7.)

Chapter 5

DIFFICULT METALS

While braze welding will work on virtually every metal with which the homeowner or farmer is familiar, there are a few metals that require either extra steps (such as preheating and postheating) or extra care to prevent changes in metal structure and resulting weakness or corrosion. Some metals also pose special hazards, as already indicated in the chapter on safety.

Lead: Avoid any work with lead or lead compounds.

Zinc: Wear proper respiratory gear, and have adequate ventilation when working with galvanized materials.

Cast Iron: In almost every case the cast iron we run into will be gray in color along a fracture line. This cast iron contains a great many impurities, but provides great strength (though ductility is generally quite poor even in an as-cast state: in other words, if dropped on a hard surface, a cast iron part will often crack quite easily, possibly even shatter). Rapid application and removal of extreme heat will cause internal stresses in the metal which will at best decrease ductility even more than normal and, at worst, shatter the pieces being joined. Thus both preheating and postheating are recommended for repairs on any cast iron part. For very large parts, this can require you to remove the part from its normal location (as, for example, an engine block), place it in a bed of charcoal, and bring the entire part up to a dull red color. Most parts (manifolds, engine heads, etc.) can be preheated with an oxy-fuel gas torch, though. Of course, all joint preparations should be made before preheating, with a great preference going to opening up a wide groove, at least 90°, along the line of the crack or fracture, no matter the type of torch manipulation used. If the groove is cut with a torch or grinder, it must be treated with an oxidizing flame to remove scattered carbon particles which will interfere with the molecular bonding of the tinning layer of braze filler metal. In general, even a very, very efficient single-gas torch will be insufficient to heat even moderate size castings for braze welding. In almost all cases an oxy-fuel gas torch is necessary.

Postheating with cast iron, after braze welding, is not an essential, but will add to joint ductility, though actually a simple retarding of the cooling action is more than enough to keep things strong: depending on the size of the piece being worked on, various materials can be used to retard cooling. For almost any size piece, one or two of the foil blankets known as space blankets do a fine job; they are easily cleaned (if you're careful not to tear them), are relatively cheap, and have other uses.

A good, active flux is essential when braze welding cast iron.

Aluminum: Aluminum becomes difficult to braze weld because of two properties that make it valuable in many other uses. First, it transfers heat very, very rapidly, making it difficult to apply enough heat to form a strong bond along thick sections. Fortunately, most aluminum we'll be working with is relatively thin in cross section. If you have to work with aluminum tubing or pipe, heat will be lost very rapidly as air flows through the center. Plug the ends for better heat retention. Use a torch tip about two sizes larger than for mild steel. Aluminum oxides are rather easily removed, as they don't extend below the surface, but the rapid reformation of new oxide coats is a problem. A very active flux is needed, as well as a good mechanical cleaning, in order to get a good bond. Preheating and postheating are not needed, though that depends in part on the filler metal used: in this case, follow the instructions of the filler rod manufacturer.

Brass: Brass is an alloy of copper and zinc. If the copper content is over 65 percent, you will need to use acetylene, though sometimes, because the melting point of the metal itself is rather low (about 1650° F.), a MAPP or propane gas torch can be used to provide sufficient heat for bonding even then. Again, a proper flux is needed, and great care must be taken to ensure that the base metals being worked do not melt, as the melting temperature of the filler metal may be very close to that of the base metal. Use a slight oxidizing flame for this metal, if an oxy-fuel gas torch is used. Preheat to about 500°F. with the torch. Heat dissipation is rapid, so a larger tip is required than you would need for mild steel. Go up about one size.

Bronze: Another copper alloy, this time with tin, bronze melts at a temperature just a few degrees lower than does brass. Flux is needed, as is a larger tip to counteract rapid heat flow. An oxidizing flame will start to melt bronze as it reaches red heat so, basically, this type of flame is a good indicator of too much heat. Again, preheat to 450° or 500°.

Stainless Steel: There is a wide variety of stainless steel around today. To fusion weld such metals usually requires more sophisticated procedures than most people care to learn (or to buy the equipment for). Thus it is often braze welded or hard soldered. Make certain the filler rod is specified for stainless steel and not just steel. Use a slight carburizing flame, depending on the filler rod's manufacturer to tell you whether you'll need 1X, 1½X, or other flame types. Use a tip about a size *smaller* than for braze welding mild steel of the same size. Flux.

In most cases, we'll be working with mild steel or cast iron when braze welding. Hard soldering will find us working with the other metals. Following basic instructions as covered in this book, along with any recommendations from the torch and filler rod manufacturers will, 99 times out of 100, result in a good, strong job. Often, you can save an engine, a plow, or some other expensive piece of gear that would have to be discarded.

ELECTRIC ARC BRAZING AND BRAZE WELDING

Throughout the book, we've emphasized the use of fuel gases or fuel gases and oxygen as heat-producing agents for braze welding and brazing. In general, these processes are considered to be gas welding methods. Not often used, arc welders, with appropriate accessories, can, with care, be used if you wish. Assuming you have previously purchased the equipment, you may prefer to use it.

In that case, you will need to purchase a twin carbon arc torch similar to that illustrated here. In general, such a torch will add less than $20 to the cost of your arc welding gear. It produces a searingly hot flame of about 9000° F. and great care in manipulation of the torch is essential. Thin sheet metal is difficult, but thicker sections work quite as easily as with oxy-fuel gas torches. Most of the basic instructions for an oxy-fuel

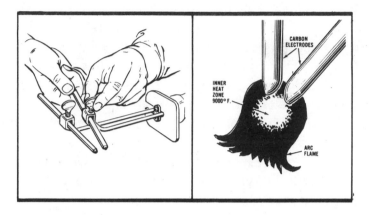

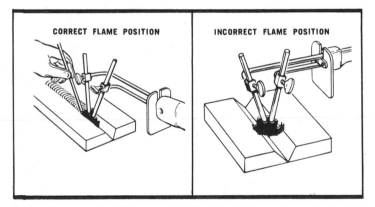

Figure 5-1. The use of twin carbon arc torch. SEARS, ROEBUCK AND CO.

gas braze weld apply, with the exception of tip selection: carbon arc torches increase capacity as the thickness of the carbon rods used to form the arc is increased. (Fig. 5-1.)

While the carbon arc can also be used for brazing, it is not too easily handled because of the extreme heat of the arc flame. Generally, it would pay even the most avid of arc welders to purchase a MAPP or propane torch for that purpose.

GLOSSARY

acetone (CH_3COCH_3). A solvent having the ability to dissolve or absorb many times its own volume of acetylene. Employed as the liquid solvent in acetylene cylinders. See **acetylene**.

acetylene (C_2H_2). A combustible, endothermic gas generated by the action of water on calcium carbide. Endothermic means that heat is absorbed by the gas when it is formed and given up in addition to the heat of combustion when burned. In the presence of pure oxygen, the acetylene flame will produce a temperature of approximately 6300° F. This flame is employed in oxyacetylene welding, cutting, flame hardening, flame cleaning, gouging, deseaming, etc.

alignment. Arrangement of parts in proper position for welding.

A.W.S. Abbreviation for American Welding Society.

backfire. Momentary retrogression or burning back of the torch flame into the torch tip. Immediately following the withdrawal of the tip from the work, the gases may be reignited by the hot work piece; otherwise the use of a lighter may be necessary.

backhand welding. That method of welding in which the torch and rod are so disposed in the vee that the torch flame points back at the completed weld, enveloping the newly deposited metal, and the rod is interposed between the torch and the weld.

base metal. Materials composing the pieces to be united by welding. Also called **parent metal**.

bead weld. A type of weld made by one passage of rod and torch.

bevel. A special preparation of metal to be welded wherein the edge is ground or cut to an angle other than 90° to the surface of the material.

blow hole. A hole or cavity formed by trapped gas or by dirt, grease, or foreign substances. If it appears at the point of junction of base metal and weld metal, it usually is caused by too rapid melting of the metal. If near the surface, it may be the result of too rapid cooling.

bond strength. In a braze-welded joint, the strength at the juncture of the braze weld metal and the base metal. This strength is dependent upon the intermolecular penetration of the brazing alloy into the surfaces of the base metal.

brass. A copper-zinc alloy, the melting point of which will vary with the analysis. A typical analysis is: copper, 63%; zinc, 37%; melting point $1670°$ F.

braze. A group of welding processes wherein coalescence is produced by heating to suitable temperatures above $800°$ F. and by using a nonferrous filler metal having a melting point below that of the base metals. The filler metal is distributed in the joint by capillary attraction.

butt joint. A welded joint between two abutting parts lying in approximately the same plane.

carburizing flame. A gas flame having the property of introducing carbon into the metal heated.

corner joint. A welded joint at the junction of two parts located approximately at right angles to each other in the form of an L.

combustion. The process of rapid oxidation or burning.

cylinder. A portable metallic container for storage and transmission of compressed gases.

deoxidized copper. Copper from which the oxygen has been removed by the addition of a deoxidizer, phosphorus or silicon. This lowers the electrical conductivity but yields a product more suitable for oxyacetylene welding.

ductility. The property of metals that enables them to be mechanically deformed without breaking when cold.

Easy-Flo. A low temperature brazing alloy nominally consisting of silver, 50.0%; copper, 15.5%; zinc, 16.5%; cadmium, 18.0%.

edge joint. A welded joint connecting the edges of two or more parallel or nearly parallel parts.

electrolytic copper. Copper refined by the electrolytic method, yielding metal of high purity (over 99.94% copper).

face of weld. The exposed surface of a weld.

ferrous metals. Those metals or alloys the principal constituent or base of which is iron.

filler metal. Material to be added in making a weld.

fillet weld. A weld of approximately triangular cross section, as used in a lap joint, tee joint, or corner joint, joining two surfaces approximately at right angles to each other.

flanged edge joint. A joint in two pieces of metal formed by flanging the edges at 90° to the plates and joining with an edge weld. See **edge joint.**

flashback. The retrogression or burning back of the flame into or beyond the mixing chamber, sometimes accompanied by a hissing or squealing sound and the characteristic smoky, sharp-pointed flame of small volume. When this occurs, immediately shut off the torch oxygen valve and then the acetylene valve.

flux. A chemical compound or mixture in powdered, paste, or liquid form whose essential function is to combine with or otherwise

render harmless those products of the welding or brazing operation which would reduce the physical properties of the deposited
metal or make the welding or brazing operation difficult or impossible.

forehand welding. That method of welding in which the torch and
rod are so disposed in the vee that the torch flame points ahead
in the direction of welding and the rod precedes the torch.

gas pocket. A cavity in a weld caused by gas inclusion.

gray iron. Pig- or cast-iron containing 3.00-3.75% carbon, a large part
of which is in the free graphitic state; gray iron is nonmalleable
at any temperature.

groove weld. A weld made by depositing filler metal in a groove between two members to be joined. See **butt joint.**

inclusion. A gas bubble or nonmetallic particle entrapped in the weld
metal as a result of improper manipulation.

joints. See **butt joint, corner joint, lap joint, tee joint,** etc.

kerf. The space from which the metal has been removed by a cutting
process.

land. The portion of the prepared edge of a part to be joined by a
groove weld, which has not been beveled or grooved. Sometimes
called **root face.**

lap joint. A welded joint in which two overlapping parts are connect-
ed, generally by means of fillet welds.

low temperature brazing. That group of the brazing processes wherein the brazing alloys employed melt in the range of about $1175°$-
$1300°$F. and the shear type (lap) joint is used.

mixing chamber. That part of the welding or cutting torch wherein gases are mixed for combustion.

multilayer welding. In oxyacetylene welding a technique in which a weld, on thick material, is made in two or more passes.

neutral flame. A gas flame wherein the portion used is neither oxidizing nor reducing.

nonferrous. Metals containing no substantial amounts of ferrite or iron; examples are copper, brass, bronze, aluminum, and lead.

oxidation. The process of oxygen combining with elements to form oxides.

oxide. A chemical compound resulting from the combination of oxygen with other elements.

oxidizing flame. A gas flame wherein the portion used has an oxidizing effect.

oxygen. A colorless, odorless gas, essential to all combustion reactions.

parent metal. See base metal.

phos-copper. A low-temperature brazing alloy consisting of copper, 93%; phosphorus, 7%.

porosity. The presence of gas pockets or inclusions.

postheating. Heat applied subsequent to welding or cutting operations.

preheating. Heat applied prior to welding or cutting operations.

psi. Standard abbreviation for "pounds per square inch."

regulator. A mechanical device for accurately controlling the pressure and flow of the gases employed in welding, cutting, braze welding, etc.

rod, welding.   Filler metal in rod or wire form, used in the gas welding processes.

root face.   See **land.**

root opening.   The separation at the root between parts to be joined by a groove weld.

root of weld.   The point at the bottom of the weld.

Sil-fos.   A low-temperature brazing alloy nominally consisting of silver, 15.0%; copper, 80.0%; phosphorus, 5.0%.

slag inclusion.   Nonmetallic material entrapped in a weld.

spelter.   A term applied to powdered brass used in making a typical brazed (lap) joint.

tack weld.   A weld used for assembly purposes only.

tee joint.   A welded joint at the junction of two parts located approximately at right angles to each other in the form of a T.

throat of fillet weld.   Actual: The distance from the root to the face of the weld.   **Theoretical:**   The distance from the root to the hypotenuse of the largest isosceles right triangle that can be inscribed within the weld cross section.

tin, to.   See **wet, to.**

toe of weld.   The junction between the face of the weld and the base metal.

torch.   A device used in gas welding or cutting for mixing and controlling the gases.

wet, to.   Free unobstructed flow of brazing alloy on the base metal.

APPENDIXES

TEMPERATURES:

**Temperatures expressed
as DEGREES CELCIUS:**

°F	°C
425	219
1050	566
1090	588
1125	607
1130	610
1400	760
1600	871
2000	1093

TEMPERATURE CONVERSION CHART

°C	°F
1000	1800
950	1700
900	1600
850	1500
800	1400
750	1300
700	
650	1200
600	1100
550	1000
500	900
450	800
400	700
350	600
300	
250	500
200	400
150	300
100	200
50	100
0	0
-50	

APPENDIX B

MELTING POINTS OF METALS

Metal or Alloy	Melting Point, °F
Aluminum, Pure	1218
Brass and Bronze	1600-1660
Copper	1981
Iron, Cast and Malleable	2300
Lead, Pure	620
Magnesium	1240
Monel	2400
Nickel	2646
Silver, Pure	1762
Steel, Hi-Carbon (0.40% to 0.70% Carbon)	2500
Steel, Medium Carbon (0.15% to 0.40% Carbon)	2600
Steel, Low Carbon (less than 0.15%)	2700
Stainless Steel, 18% Chromium, 8% Nickel	2550
Titanium	3270
Tungsten	6152
Zinc, Cast or Rolled	786

SUGGESTED GAS WELDING PROCEDURES

BASE METAL	GAS PROCESS*	FLAME TYPE
Aluminum Alloys 2S & 3S	Welding	Neutral
	Brazing	Neutral
Aluminum Alloys 52S, 53S, 61S, & 63S	Welding	Neutral
	Brazing	Neutral
Brass, Red	Welding	Oxidizing
Brass, Yellow	Welding	Oxidizing
Bronze, Aluminum (Below 5% Al.)	Welding	Slightly Carburizing
Bronze, Phosphor	Welding	Neutral
	Braze Welding	Oxidizing
Copper, Beryllium	Brazing	Slightly Carburizing
Copper, Deoxidized & Copper, Electrolytic	Welding	Slightly Oxidizing or Neutral
	Braze Welding	Slightly Oxidizing
Muntz Metal	Welding	Slightly Oxidizing
Nickel & High Nickel Alloys	Welding	Neutral or Slightly Carburizing
Nickel Silver	Braze Welding	Neutral

*All of the metals in the table with the exception of the Aluminum alloys can be brazed using either the silver or copper base brazing alloys. The melting point and composition of the base metal will determine the correct brazing alloy.

FILLER METAL	FLUX	REMARKS
Airco No. 25	Airco Napolitan	
Airco No. 26 or 718	Airco Elite	
Airco No. 26	Airco Napolitan	
Airco 718	Airco Elite	Recommend etch with acid before brazing
Airco No. 20, 22, or 27**	Airco Hi-Test or Marvel	
Airco No. 20	Airco Hi-Test or Marvel	
Match base metal	Special Flux	Weld as continuously as possible. Don't weld Aluminum Bronze with above 5% Al.
Match base metal (Grade E phosphor bronze)	Airco Hi-Test or Marvel	
Airco No. 20, 22, or 27	Airco Hi-Test or Marvel	
Aircosil 50, 45, 35, or 3	Aircosil Flux	Only process recommended
Airco No. 23A	Airco Marvel	
Airco No. 20, 22, or 27	Airco Marvel	
Airco No. 20, 22, or 27	Airco Hi-Test or Marvel	Airco No. 22 espepecially recommended for manganese bronze
Match base metal	Manufacturer's recommendation	
Airco No. 21	Airco Hi-Test or Marvel	

**Note:* Certain corrosive conditions do not permit the use of high zinc brass filler metals; for those cases weld with silicon copper rods.*

APPENDIX D | SUGGESTED GAS WELDING PROCEDURES

BASE METAL	GAS PROCESS*	FLAME TYPE	
Cast Iron, Gray	Welding	Neutral	
	Braze Welding	Neutral	
Cast Iron, Malleable	Braze Welding	Slightly Oxidizing	
Cast Iron, Moly	Welding	Neutral	
Galvanized Iron or Steel	Braze Welding		
Steel, Cast	Welding	Slightly Carburizing	
	Braze Welding	Neutral	
Steel, High Carbon (0.45% and up)	Welding	Carburizing	
Steel, Low Carbon (up to 0.30%)	Welding	Slightly Carburizing	
Steel, Medium Carbon (0.30-0.45%)	Welding	Slightly Carburizing	
Steel, Miscellaneous Alloy	Welding	Slightly Carburizing	
Steel, Stainless	Welding	Slightly Carburizing	
Wrought Iron	Welding	Neutral	
	Braze Welding	Neutral	

*All of these metals can be brazed using the Easy-Flo's or Airco Nos. 20, 22, or 27.

ROD	FLUX	REMARKS
Airco No. 9	Airco Atlas	Preheat (900° F. min.)
Airco No. 20	Airco Hi-Test, Hi-Bond, or Marvel	Slight preheat
Airco No. 20	Airco Hi-Test, Hi-Bond, or Marvel	
Airco No. 10	Airco Atlas	Preheat (900° F. min.) Postheat (1100° F. min.)
Airco No. 22 or No. 27	Airco Hi-Test	Provide proper ventilation to get rid of zinc fumes
Airco No. 4	None	
Airco No. 22 or No. 27	Airco Hi-Test, Hi-Bond, or Marvel	
Airco No. 1 or No. 4**	None	
Airco No. 1, 4, or 7	None	
Airco No. 1 or No. 4	None	
Special rods	None	
Rod matching base metal	Airco Stainless Steel Flux	
Airco No. 1, 4, or 7	None	
Airco No. 20, 22, or 27	Airco Hi-Test Hi-Bond, or Marvel	

**Note: Satisfactory for some conditions, but other conditions will require a high carbon filler metal.*

APPENDIX E

VARIATIONS IN OXYGEN CYLINDER PRESSURES
WITH TEMPERATURE CHANGES

Gauge pressures indicated for varying temperature conditions on a full cylinder initially charged to 2200 psi at 70° F. Values identical for 244 cu. ft. and 122 cu. ft. cylinder.

Temperature Degrees F.	Pressure psi approx.	Temperature Degrees F.	Pressure psi approx.
120	2500	30	1960
100	2380	20	1900
80	2260	10	1840
70	2200	0	1780
60	2140	−10	1720
50	2080	−20	1660
40	2020		

APPENDIX F

OXYGEN CYLINDER CONTENT

Indicated by Gauge Pressure at 70° F. 244 cu. ft. Cylinder

Gauge Pressure psi	Content cu. ft.	Gauge Pressure psi	Content cu. ft.
190	20	1200	130
285	30	1285	140
380	40	1375	150
475	50	1465	160
565	60	1550	170
655	70	1640	180
745	80	1730	190
840	90	1820	200
930	100	1910	210
1020	110	2000	220
1110	120	2090	230
		2200	244

122-cu. ft. cylinder content one-half above volumes.

INDEX

acetone, 18, 85
acetylene, 10, 11, 17, 21, 33, 82, 85
acetylene hose, 43
acetylene regulator, 43
Airco Tote-Weld torch, 17, 35, 42
aluminum, 9, 25, 47, 55, 61, 70, 82
argon, 24
asbestos millboard, 10, 13, 16

backfire, 42, 48-49, 85
backhand welding, 57-58, 61, 66, 85
base metal, 10, 51, 53, 55, 64, 66, 82, 85
bead weld, 52, 85
Bernzomatic torch, 10, 31, 75
bevel, 53, 55, 57, 58, 85
blow hole, 86
bond strength, 86
brass, 51, 82, 86
braze welding, 9, 10, 13, 14, 17, 19, 26, 35, 37, 38, 45, 51-67, 81, 82
  flux, 63-65
  joint design, 54-57
  joint preparation, 53-56
  procedure, 65-67
  temperature, 51, 53
  torch adjustment, 59-63
  torch handling, 57-58
brazing, 9, 10, 11, 13, 14, 19, 24-25, 35, 69-79

container brazing, 24-25
  flux, 72-74, 76, 79
  heat application, 75-79
  joint design, 70-74
  surface preparation, 70-72
  temperature, 69, 73
British thermal unit (BTU), 33
bronze, 51, 83
burns, 18
burnthrough, 51, 55
butt joint, 52, 66, 72, 86
butt weld, 55, 58

cadmium, 25
candle, 17, 35
capillary attraction, 51, 69, 72
carbon arc torch, 83-84
carburizing flame, 45, 46, 66, 83, 86
cast iron, 9, 26, 51, 53, 59, 64, 65, 66, 67, 81, 82, 83
charcoal, 81
check valve, 39-42
chromium, 25
Cleanweld Solidox torch, 17, 35
clothing, 14-16
cold chisel, 53
color coding, 21
combustible surfaces, 16
container brazing, 24-25
copper, 25, 47, 53, 82, 83
corner joint, 52, 86
corner weld, 55

DRAKE HOME CRAFTSMAN'S BOOKS

AN INTRODUCTION TO WORKING WITH METAL
This definitive book for the beginning metalworker is
arranged in six sections with easy-to-follow line drawings
and instructions.
$4.95 paper / 128 pages / 6 x 9 / ISBN: 0-8473-1652-1

METAL WORKING   Oscar Almeida
Aimed especially for the home craftsman, this comprehen-
sive work covers such metalwork as design, fitting, alloys,
joining casting, grinding, and polishing.
$4.95 paper / 275 pages / 6 x 9 / ISBN: 0-8473-1172-4

SOLDERING AND WELDING   B. M. Allen
All aspects of soldering and welding are discussed, with a
scientific approach, for both the industrial artisan and the
home craftsman.
$4.95 paper / 128 pages / 6 x 9 / ISBN: 0-8473-1120-1

THE COMPLETE HANDYMAN   Ed. Charles H. Hayward
How to make your own home repairs: painting and paper-
hanging, brickwork, glass and glazing, concrete, roofing,
floors, doors, windows, and much more.
$5.95 paper / 352 pages / 7 x 10 / ISBN: 0-8473-1341-7

INTRODUCTION TO FURNITURE MAKING
John R. Trussell
How to design and make your own furniture, including
discussion of timber, designs, joints, special constructions,
drawers, laminating, glues, and finishing.
$4.95 paper / 144 pages / 6 x 9 / ISBN: 0-8473-1339-5

PRACTICAL UPHOLSTERY   C. Howes
Principles and practical applications of the craft, as well
as a whole range of designs for all types of furniture, both
traditional and contemporary.
$4.95 paper / 128 pages / 6½ x 8¼ / ISBN: 0-8473-1690-4

ADHESIVE BONDING OF WOOD   M. L. Selbo
A complete guide to glue products, discussing wood properties, construction types, preparing wood for gluing, bonding
processes, gluing operations, etc.
$4.95 paper / 128 pages / 6 x 9 / ISBN: 0-8473-1666-1

CARPENTRY FOR BUILDERS   A. B. Amary
Shoring, timbering, centers for arches, gantries, form work
for concrete, roofs, geometry, and the steel square in roofing are among the woodworking jobs covered.
$4.95 paper / 352 pages / 6 x 9 / ISBN: 0-8473-1159-7

COMPLETE CARPENTER'S HANDBOOK
Carpentry construction techniques for the amateur craftsman, covering blueprint reading, tools, lumber, frame construction, joints and splices, roofing, etc.
$4.95 paper / 160 pages / 8¼ x 11 / ISBN: 0-8473-1117-1

HOW TO BUILD WOODFRAME HOUSES
A handbook for beginners, this book covers all facets of
woodhouse construction: site, foundation, wall sheathing,
fireplaces, and even maintenance and repair.
$4.95 paper / 232 pages / 8¼ x 11 / ISBN: 0-8473-1337-9

MAKING TOYS IN WOOD   Charles H. Hayward
Invaluable for both the family woodworker and the crafts-
person, this book covers the ideal material for children's
toys, and designs and plans for popular and durable projects.
$4.95 paper / 168 pages / 5½ x 8¼ / ISBN: 0-87749-704-4

MODERN WOOD TURNING   Gordon Stokes
A complete guide to wood turning for every home woodworker, student, and shop instructor.
$4.95 paper / 128 pages / 6½ x 8½ / ISBN: 0-8473-1306-9

PRACTICAL WOODWORK   Charles H. Hayward
The basic essentials of woodwork and furniture making
are explained, from handling tools through finishing. Working drawings and cutting lists included.
$4.95 paper / 192 pages / 5½ x 8¼ / ISBN: 0-8473-1683-1

STAINING AND POLISHING   Charles H. Hayward
Wood surfaces, techniques, coloring, staining, and finishing;
developed into exhibitions of artisanship.
$4.95 paper / 218 pages / 5½ x 8¼ / ISBN: 0-87749-708-7

TOOLS FOR WOODWORK   Charles H. Hayward
Every tool needed by the craftsman is described, as well as
possible uses and tips for proper care.
$4.95 paper / 128 pages / 6 x 9 / ISBN: 0-8473-1338-7

WOOD FINISHING   F. N. Vanderwalker
A definitive work on wood finishing as it is used for decora-
tion and trim.
$4.95 paper / 416 pages / 6 x 9 / ISBN: 0-87749-811-3

WOODWORK JOINTS   Charles H. Hayward
For both the professional in training and the home carpenter; explicit instructions for commonly used woodwork
joints and variations, with many diagrams.
$4.95 paper / 192 pages / 5½ x 8¼ / ISBN: 0-87749-707-9

WOODWORKING AND FURNITURE MAKING
G. W. Endacott
A complete guide for the home carpenter — tools, machin-
ery, woods, materials, detailed designs.
$4.95 paper / 228 pages / 6 x 9 / ISBN: 0-8473-1161-3